Executive Teams in Research-Based Spin-Off Companies

Rigo Tietz

Executive Teams in Research-Based Spin-Off Companies

An Empirical Analysis of Executive Team Characteristics, Strategy, and Performance

Foreword by Prof. Dr. Michael Schefczyk

 Springer Gabler

Rigo Tietz
St.Gallen, Switzerland

University of Technology Dresden, 2012

The study was funded by the Free State of Saxony
and the European Union within the programme
"Landesinnovationspromotionen".

Europa fördert Sachsen.

Gefördert aus Mitteln
der Europäischen Union

Europäischer Sozialfonds

ISBN 978-3-658-01214-4   ISBN 978-3-658-01215-1 (eBook)
DOI 10.1007/978-3-658-01215-1

The Deutsche Nationalbibliothek lists this publication in the Deutsche Nationalbibliografie;
detailed bibliographic data are available in the Internet at http://dnb.d-nb.de.

Library of Congress Control Number: 2013935229

Printed on acid-free paper

Springer Gabler is a brand of Springer DE.
Springer DE is part of Springer Science+Business Media.
www.springer-gabler.de

Foreword

The doctoral thesis of Rigo Tietz focuses on management teams of research based young enterprises which can be characterized as spinoffs. The motivation for the thesis derives from the need to make as many of such spinoffs successful as plainly possible. Research based spinoffs are highly beneficial for their founders in terms of a professional perspective, beneficial for the transferring institutions by creating an environment of dynamic firms and beneficial for society via promoting technology transfer and innovation. At the same time, substantial hurdles need to be overcome. Such hurdles include a sufficiently market oriented mindset of the founders and/or managers as well as the motivation of transferring institutions to let most promising staff go while keeping them within a useful network. Most of these hurdles are coined by people issues such as motivation, changing roles and required skills. Thus, it makes most sense to emphasize characteristics of management teams of such spinoffs.

Tietz utilizes the resource based view and human capital theory to derive so called "upper echelons" characteristics of the management teams required in the situation of a young enterprise spinning off of a research institution. He looks for a direct impact of such characteristics on the success of the ventures as well as a – conceivably more realistic – indirect influence on success which is moderated particularly by the corporate strategy chosen.

Tietz uses established theory and literature to build a substantial set of hypotheses and research questions which he then tests using a data set of 193 spinoffs from Germany. Indeed Tietz finds that the performance of spinoffs typically does not predominantly depend directly on characteristics of the management teams. Among the direct effects, at best prior professional experience plays a mildly significant role. Much more important, however, is the effect of characteristics of the management team on performance in the context of the strategic setting. The moderating effect of corporate strategy is well demonstrated in particular via two structural equation models. In summary, neither a strong management team nor a suitable strategic plan is sufficient to succeed. Both aspects must meet high standards and fit well in the light of the particular setting of each spinoff.

To summarise, the doctoral thesis makes an important contribution towards our understanding of a combination of factors that contribute to making research based spinoffs successful. The relevance of the issues, the soundness of hypotheses, the data set and the analysis are key strengths of this work. I recommend this excellent study to practitioners at research institutions with the potential to create spinoffs, private and public financiers of spinoffs and the scientific community as an inspiring and enlightening piece of literature!

Prof. Dr. Michael Schefczyk

Table of contents

List of figures

List of tables

List of abbreviations

&	And
AVE	Average variance extracted
CEO	Chief executive officer
e.g.	Exemplum grati, for example
EUR	Euro
et al.	Et alii, and others
etc.	Et cetera, and so on
excl.	Excluded
df	Degrees of freedom
FTE	Full-time equivalent
IPO	Initial public offering
KS	Kolmogorov-Smirnov-test
MIT	Massachusetts Institute of Technology
NACE	Nomenclature générale des activités dans les Communautés Européennés
No.	Number
n.s.	Not significant
NTBV	New technology-based venture
p.	Page
PDF	Portable document format
Ph.D.	Doctor of Philosophy
pp.	Pages
R^2	Squared multiple correlation
R&D	Research and development
ROA	Return on assets
ROI	Return on investment
SD	Standard deviation
TMT	Top management team
TTO	Technology transfer office
UK	United Kingdom
US	United States of America
VC	Venture capital

1   Introduction

Executive teams in research-based spin-off companies face specific challenges in establishing profitable ventures. On one hand, spin-offs are based on sophisticated knowledge and technology transferred from the parent institution to the new firm. On the other hand, executives often lack entrepreneurial and management skills when they change from the research environment to the new business. Generally, academic entrepreneurship and research-based spin-offs in particular have become important entities for researchers and practitioners. This chapter provides a brief introduction to the topic and outlines its relevance, the motivation and objective of this study, and the thesis structure.

1.1   Relevance of the subject

New technology-based ventures can contribute to local and regional development through job creation and innovation (Bollinger, Hope, and Utterback, 1983; Dorfman, 1983; Jones-Evans, and Klofsten, 1999; Steffensen, Rogers, and Speakman, 2000; Shane, 2004; Venkataraman, 2004; Fontes, 2005). They can help create industry clusters and foster diversification to strengthen the local economy. In this context, research-based spin-off companies are a special subgroup of new technology-based ventures (Mustar, Renault et al., 2006). Spin-offs that have emerged from the academic environment transform and commercially exploit knowledge and technology developed in a university or non-university research institute. The importance of new technology-based ventures is exemplified by the high-tech areas near the top-tier US universities, such as Silicon Valley, which is near Stanford, and Route 128, which is near MIT. Prominent examples such as Genentech or Google demonstrate how technology-based firms can dramatically shape their surroundings. Without a doubt, some technology-based ventures have changed established industry sectors or even created new ones by commercialising their technologies. Researchers interested in this phenomenon have sought to measure the economic contribution of new technology-based ventures and research-based spin-offs. Roberts and Malone (1996) stated that spin-offs from MIT contribute 10 billion US dollars per year to the economy and have created 300,000 jobs. Studies in the European context have come to similar conclusions. In a comprehensive study, Egeln, Gottschalk et al. (2002) found that spin-offs were responsible for 34,000 jobs per year in Germany from 1996 to 2000. McQueen and Wallmark (1991) estimated that spin-offs from Chalmers University in Sweden contributed more than 100 million US dollars per year to the Swedish economy from 1964 to 1991. Blair and Hitchens (1998) reported that research-based spin-off companies develop more innovative products and services than other new technology-based ventures. Along the same line, Shane (2004) noted that spin-offs commercialise inventions "which are too uncertain or early for established companies to pursue" (Shane, 2004, p. 105).

Because academic spin-offs are assumed to have a large effect on the economy, politicians and practitioners have introduced support programmes, built infrastructure, and designed investment schemes to foster the commercialisation of research results. In this respect, the duties and responsibilities of universities and research organisations have changed. For decades, the primary activities of universities have been teaching and basic research. Today, the so-called *third mission* consists of additional activities that aim to transfer technology and research results to the business environment. Universities and non-university research institutes have generally become more active in this area (Shane, 2002; Nerkar and Shane, 2003). Chiesa and Piccaluga (2000) outlined six factors that are responsible for the increased interest towards technology transfer activities: (1) decreased public funding forces universities to search for alternative financial sources; (2) the expectations of research organisations in terms of knowledge production and distribution have changed; (3) competition among universities has intensified; (4) pressure in terms of social accountability has increased; (5) several recently emerged industry sectors, such as biotechnology, can directly apply research findings; and (6) the incentive structure for qualified researchers has changed due to increased competition in the knowledge market.

Spin-off companies are an alternative technology transfer method to patenting, licensing, contract research, and other activities (Rogers and Hall, 1999; Debackere, 2000; Klofsten and Jones-Evans, 2000; Agrawal, 2001). Academic entrepreneurship deals with the creation of new companies to exploit scientific knowledge and has become an important field. Nevertheless, whether spin-offs lead to higher financial returns than other technology transfer instruments remains controversial. Bray and Lee (2000) stated that spin-offs generate more financial returns than licenses do. In contrast, Lerner (2005) found that academic spin-offs make only a very limited direct financial contribution. However, not all positive effects can be measured financially. Because spin-offs are usually established near the parent institution, they can provide jobs for graduates and foster local economic development (Di Gregorio and Shane, 2003). Debackere (2000) indicated that the time between knowledge creation and application is shorter for spin-offs than for other commercialisation mechanisms. Thus, new technology and knowledge are transferred to the market more quickly (Chrisman, Hynes, and Fraser, 1995). However, this development is not without its drawbacks. Some scholars have recognised a shift from basic research to more applied science and a limitation of academic freedom (Rogers and Hall, 1999; Chiesa and Piccaluga, 2000; Rothaermel, Agung, and Jiang, 2007). Researchers have also discussed whether a stronger emphasis on patenting research findings to protect intellectual property, which is often required to attract external investors, such as venture capitalists, for the commercialisation process, diminishes publishing activities. Van Looy, Callaert, and Debackere (2006) concluded that researchers' entrepreneurial activities enhance rather than diminish their publication records.

These changes to the academic and research system demand closer interaction between government, industry, and academic institutions in a so-called *triple helix model* (Etzkowitz, 1998; Leydesdorff, 2000; Etzkowitz and Klofsten, 2005). In addition, the commercialisation process in Europe is less effective than in the United States (Klofsten and Jones-Evans, 2000; Goldfarb and Henrekson, 2003). Although European universities and research organisations produce more academic output in categories such as publication records, they lag behind in transferring this knowledge to the business environment to promote innovation and economic development. This phenomenon is known as the *European paradox*. The legal system is said to be a key cause of this disparity between Europe and the US. An effective technology transfer process requires an appropriate legislative framework (Mowery, Nelson et al., 2001). In 1980, the US government introduced the Bayh-Dole Act, which stated that research organisations, not individual scientists, own intellectual property rights and have the first right to exploit them commercially (Agrawal, 2001; Mowery, Nelson et al., 2001). Meanwhile, most European countries have introduced similar legislative frameworks to connect intellectual property to academic institutions. But it will still take time to catch up in terms of spin-off venturing as an effective instrument in the technology transfer process.

In summary, academic entrepreneurship with a focus on creating new ventures based on publicly financed research has only recently become a concern for most European authorities (Stankiewicz, 1994; Mustar, 1997). Nowadays, new technology-based ventures and research-based spin-off companies are an important factor for regional development, job creation, and innovation. In terms of the effectiveness of technology transfer mechanisms and commercialisation processes, European countries still lag behind the US. However, there has been substantial development in this field recently, with researchers identifying factors contributing to the successful creation and development of academic spin-offs. These factors include considerations on political, environmental, and organisational levels.

1.2   Study objective

Besides practitioners' growing interest in exploiting and commercialising research results by creating new ventures, scholars have also shown increased interest in the field of academic entrepreneurship, and spin-offs in particular, as an emerging research domain (Shane, 2004; Wright, Clarysse et al., 2007). Recent studies have tried to identify success factors and major challenges for spin-off creation and growth (Rothaermel, Agung, and Jiang, 2007). Success factors usually include the role of the parent institution in terms of intellectual property policies or support strategies and different resources, such as technology endowments, access to financial support and funding, or human capital (e.g., Shane and Stuart, 2002; Di Gregorio and Shane, 2003; Lockett, Wright, and Franklin, 2003). Besides the environment and the characteristics of the entrepreneurs, Hofer and Sandberg (1987) and Keely and Roure (1990) mentioned strategic orientation as a key success factor for new technology-based ventures. Schröder (2008) found significant links between several dimensions of the strategic orientation of new technology-based ventures and firm performance. Based on an empirical study of 149 university spin-offs in Germany, Walter, Auer, and Ritter (2006) noted that entrepreneurial orientation as an aspect of a firm´s strategic orientation affects organisational performance. Riesenhuber (2008) empirically confirmed the positive impact of an entrepreneurial orientation on performance, measured by firm growth, in the context of academic spin-offs.

On one hand, research-based spin-off companies are a subgroup of new technology-based ventures with certain shared characteristics. Thus, in the venture creation and early development phase, they face such liabilities as newness, size, adolescence, uncertainty, and owner dependence (Stinchcombe, 1965; Aldrich and Auster, 1986; Brüderl and Schüssler, 1990; Shane and Stuart, 2002). On the other hand, academic spin-offs have several specific attributes, which can imply additional barriers to establishing a successful and sustainable venture. Spin-offs often start in a very early stage of the venture development process, sometimes even without a market-ready product or portfolio, when the venture is still in a research phase (Vohora, Wright, and Lockett, 2004). Unlike most traditional start-ups, academic spin-offs are based on scientific knowledge and technology. Therefore, they are usually confronted with a high level of uncertainty because the future development of the business is largely unpredictable at the time of venture creation.

Besides the transfer of knowledge and technology from the parent institute to the new venture, the transfer of people differs substantially between typical start-ups and academic spin-offs. The entrepreneurial team of a research-based spin-off company often includes former academic staff, who generally only have experience in a non-commercial scientific environment. The literature defines academic or technology entrepreneurs as spin-off founders and managers who have been involved in the research projects on which the new venture is based (Franklin, Wright, and Lockett, 2001). They often have a scientific background with an education in a technology-oriented field or natural science, and their practical work experience is usually limited to working in the laboratory or research institute. Academic entrepreneurs tend to emphasise technical aspects rather than business-related topics, such as marketing and sales or finance and accounting (Radosevich, 1995; Meyer, 2003; Heirman and Clarysse, 2004). Thus, they generally lack entrepreneurial and management knowledge and experience. Radosevich (1995) suggested that experienced and knowledgeable managers who have not worked in the parent institute may enhance the venture development process. This approach is also called surrogate entrepreneurship. Furthermore, technology-oriented companies tend to be founded by teams rather than individuals (Roberts, 1991a). Entrepreneurial teams in academic spin-offs tend to be more homogeneous in terms of their educational background and professional work experience than are teams in other start-ups (Ensley and Hmieleski, 2005).

These observations mark the starting point of this study, which examines executive teams in research-based spin-off companies. Most studies of the effects of human capital in the entrepreneurship context use the resource-based view of the firm (Barney, 1991) to predict outcomes on the organisational level, such as firm performance. However, this approach has several limitations and shortcomings. On one hand, the resource-based view is critically discussed because of the underlying tautological and self-verifying reasoning (Priem and Butler, 2001a,b; Powell, 2001). On the other hand, human capital characteristics such as experience and knowledge are generally considered as resources, although they could constrain venture development in certain situations (Cannella, Finkelstein, and Hambrick, 2008). The strategic management literature has discussed the impact of top management team characteristics on organisational outcomes, including both performance-related aspects and firm strategy. Hambrick and Mason (1984) introduced the upper echelons perspective as a theoretical framework for a large body of studies. Nevertheless, most prior research in this area has focused on management teams in large corporations rather than on small or new ventures in the entrepreneurship context.

Based on the previous discussion, this study connects the entrepreneurship literature with the strategic management field and analyses the impact of executive teams on strategy and performance in the context of research-based spin-off companies. Potential antecedents and consequences of strategic processes will be examined. The study design considers the specific characteristics of executive teams in academic spin-offs and, thus, focuses on aspects such as prior professional experience and team composition. Furthermore, the analysis adopts the theoretical lens of the upper echelons framework and links this perspective to other theories, such as the resource-based view of the firm. The research approach includes a confirmatory part that postulates several hypotheses based on theory and previous empirical findings. When the available theoretical or empirical background is insufficient, research questions are developed to address certain relationships between team characteristics and organisational outcomes. This part of the research has an explorative character.

1.3   Structure of the thesis

This study is structured as follows. The next chapter provides a brief overview of the entrepreneurship literature in the field of entrepreneurial teams in academic spin-offs. First, different definitions and typologies are discussed to clarify the boundaries of the target group of this study. Some specific characteristics of research-based spin-off companies in contrast to traditional start-ups are outlined as well. The following section summarises prior research in the field of academic entrepreneurship and delineates the scope and theoretical background of the studies. Entrepreneurial teams are discussed in the context of success factors of research-based spin-offs that are thought to have an essential impact on venture development and performance. Finally, the chapter identifies the research gap and formulates the main research question of the study.

The third chapter describes the theoretical framework for the empirical analysis. It refers to the emerging field of strategic entrepreneurship, which combines the literature on entrepreneurship and the literature on strategic management. In this context, important strategic processes for academic spin-offs are outlined and positioned in the areas of strategy content and strategy process research. The subsequent section discusses the literature on strategic decision making and distinguishes between formal and informal strategy approaches. To link the characteristics of the management team with strategic decision making on the organisational level and firm performance, the upper echelons theory is introduced as the central framework for the study. The final section of the chapter summarises the intended contributions of the analysis.

In chapter 4, the research model is outlined formulating hypotheses and research questions for the examined relationships. Whereas the hypotheses are based on theory and empirical evidence, the research questions address those links for which prior studies have provided evidence for a significant relationship but a strong theoretical foundation is missing. The upper echelons perspective provides the framework for the empirical analysis and is complemented by additional theoretical approaches, such as the resource-based view, human capital theory, effectuation and causation theory, stage-based models, and life-cycle theory. Additionally, this section refers to prior empirical studies in the context of academic spin-offs, new technology-based ventures and small and young firms where relevant.

The fifth chapter explains the data collection process for the empirical analysis. The first part presents the procedure and findings of the preliminary study to describe the overall spin-off population. This survey was sent to technology transfer officers and the directors of universities and non-university research institutes in several European countries to identify spin-offs from their organisations. The second part of the chapter outlines the main survey of this study that addressed managing directors of a sample of German research-based spin-off companies. Finally, chapter 5 discusses the sample selection process and the survey tools and provides a statistical overview of the group of respondents compared with the entire population and the extracted sample.

Chapter 6 provides an overview of the measurement models that were applied in the empirical analysis. This overview encompasses several variables that can be observed and measured directly, including executive team attributes such as prior experience or formal qualifications. In contrast, the dimensions of the strategic orientation of the firm were operationalised by latent variables because they can only be estimated indirectly. Different constructs reflecting formal and informal approaches to strategic decision making and dimensions of an entrepreneurial strategic posture were conceptualised. Firm performance was assessed by two distinct dimensions: growth and profitability. Furthermore, several firm characteristics and environmental uncertainty were measured. The final section of the chapter highlights important aspects of the data analysis.

The seventh chapter describes the results concerning the analysed relationships between management team characteristics and firm performance, between strategic processes and performance, and between team characteristics and strategic decision making. Additionally, several influence factors on the firm and environmental levels predicting strategic processes are examined. The first step of the analyses focuses on each link separately and discusses the goodness-of-fit indices for the models. The following section combines several significant variables in an integrated model to demonstrate the impacts of certain factors in combination with each other.

Finally, chapter 8 summarises the findings and discusses the proposed hypotheses and research questions. In addition, implications for future research in the fields of academic entrepreneurship and strategic management are described. Potential implications for politicians, academic institutions, investors, and research-based spin-off companies are discussed as well. The last section of the final chapter outlines limitations and summarises the most important conclusions. Figure 1.1 shows the structure of the study.

Figure 1.1: Structure of the study

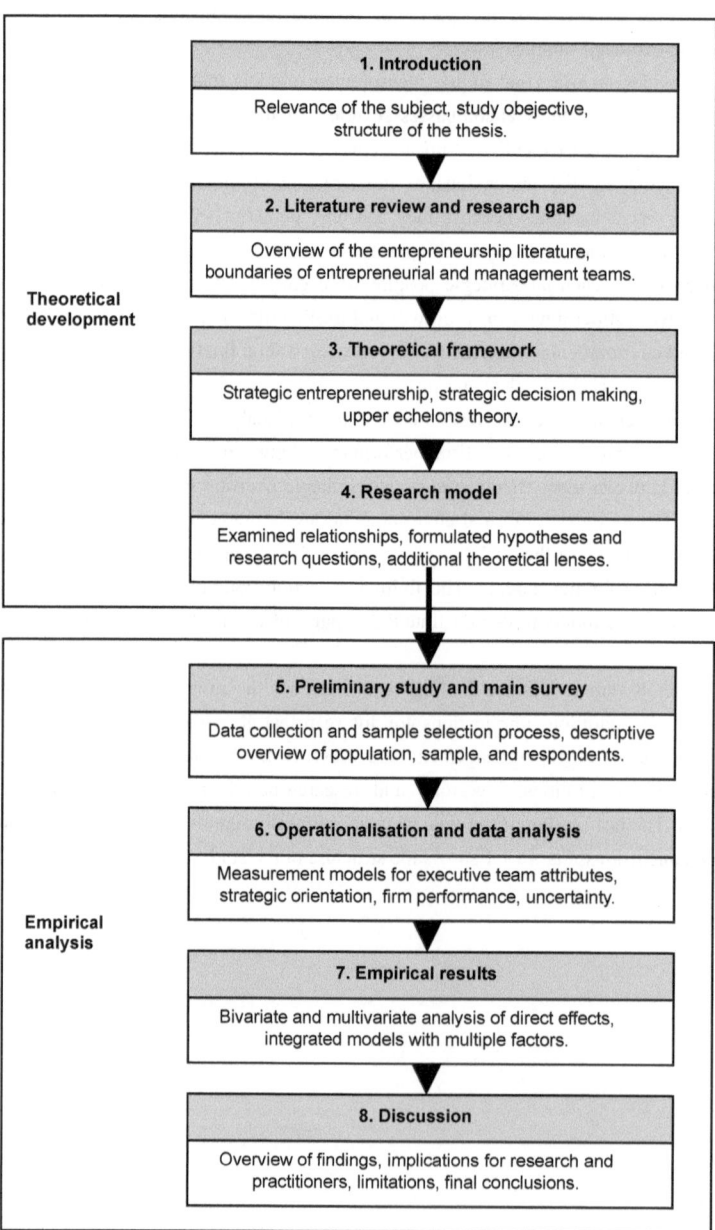

2 Literature review and research gap

The following chapter introduces the literature of academic spin-offs, covering several aspects. In the first section, different definitions of spin-off companies are presented to outline the boundaries and highlight several important differences between spin-offs and traditional start-up companies. In this context, certain existing typologies are also briefly described. On this basis, the target group of the underlying study is specified. The next step provides an overview of prior research in the field of academic entrepreneurship and outlines the scope and theoretical approaches of the studies. The following section discusses entrepreneurial and management teams in research-based spin-off companies, thus determining the central focus of this study. The main point of interest draws upon those factors, also referred to as success factors, which influence outcomes at the organisational level, such as performance. This chapter concludes by identifying the research gap and formulating the research question as the starting point for further analysis.

2.1 Overview of existing definitions

In prior studies, research-based spin-off companies have been defined as new ventures based on the transfer of technology or academic knowledge developed by public research organisations with a focus on the commercialisation of the research results. In this context, spin-offs are discussed from various perspectives, that is, from the university or firm level, for instance. Despite the increasing attention paid to the spin-off process and activities, however, most authors do not give a strict and clear definition of a research-based spin-off, thus making it difficult to compare the results of different studies (Pirnay, Surlemont, and Nlemvo, 2003). Table 2.1 provides an overview of the definitions of the spin-off phenomenon and indicates how authors clarify the concept in several ways, including similarities and differences. Researchers have used a broad variety of terms, such as spin-off or spin-out companies; university, academic, and research-based spin-offs; new technology-based ventures or high-tech start-ups, thus demonstrating the plurality of definitions in the field.[1]

---

[1] Discussed in more detail in De Cleyn and Braet (2009), referring to: Kazanjian (1988), Dahlstrand (1997), Feldman and Klofsten (2000), Di Gregorio and Shane (2003), Jagersma and van Gorp (2003), Kakati (2003), Meyer (2003), O'Shea, Allen et al. (2004), Vohora, Wright, and Lockett (2004), Aspelund, Berg-Utby, and Skjevdal (2005), Lockett, Siegel et al. (2005), Aaboen, Lindelöf et al. (2006), Mustar, Renault et al. (2006)

Table 2.1: Definitions of academic spin-off companies

Authors	Year	Definition
McQueen and Wallmark	1982	"... in order to be classified as a university spin-off, three criteria have to be met: (1) the company founder or founders have to come from a university (faculty staff or student); (2) the activity of the company has to be based on technical ideas generated in the university environment; and (3) the transfer from the university to the company has to be direct and not via an intermediate employment somewhere" (p. 307).
Smilor, Gibson, and Dietrich	1990	"a company that is founded (1) by a faculty member, staff member, or student who left the university to start a company or who started the company while still affiliated with the university; and/or (2) around a technology or technology-based idea-developed within the university" (p. 63).
Weatherston	1995	"... an academic spin-off can be described as a business venture which is initiated, or become commercially active, with the academic entrepreneur playing a key role in any or all of the planning, initial establishment, or subsequent management phases" (p. 1).
Dahlstrand	1997	"An entrepreneurial spin-off arises when an entrepreneur leaves a company to start a firm of his own. To be a spin-off, this must also include the transfer of some rights, e.g., assets or knowledge, from the existing legal body to the new firm or body" (p. 2).
Carayannis, Elias et al.	1998	"...a new company formed by individuals who were former employees of a parent organization (the university), around a core technology that originated at a parent organization and that was transferred to the new company" (p. 1).
Bellini, Capaldo et al.	1999	"... academic spin-offs are companies founded by university teachers, researchers, or students and graduates in order to commercially exploit the results of the research in which they might have been involved at the university ... the commercial exploitation of scientific and technological knowledge is realised by university scientists (teachers or researchers), students and graduates" (p. 2).
Rappert, Webster, and Charles	1999	"University spin-offs are firms whose products or services develop out of technology-based ideas or scientific/technical know-how generated in a university setting by a member of faculty, staff or student who founded (or co-founded with others) the firm" (p. 874).
Clarysse, Heirman, and Degroof	2000	"... Research-based spin-offs are defined as new companies set up by a host institute (university, technical school, public/private R&D department) to transfer and commercialize inventions resulting from the R&D efforts of the departments" (p. 546).

Table 2.1: Definitions of academic spin-off companies (continued)

Authors	Year	Definition
Klofsten and Jones-Evans	2000	"... formation of new firm or organisation to exploit the results of the university research" (p. 300).
Steffensen, Rogers, and Speakman	2000	"A spin-off is a new company that is formed (1) by individuals who were former employees of a parent organization, and (2) a core technology that is transferred from the parent organization" (p. 97).
Nicolaou and Birley	2003	"Spinouts involve: (1) the transfer of a core technology from an academic institution into a new company and (2) the founding member(s) may include the inventor academic(s) who may or may not be currently affiliated with the academic institution" (p. 333).
Clarysse and Moray	2004	"A common two-dimensional definition of a research based spin-off is a new company that is formed (1) by a faculty member, staff member or student who left university to found the company or started the company while still affiliated with the university; and/or (2) a core technology (or idea) that is transferred from the parent organisation" (p. 59).
Hindle and Yencken	2004	"Direct research spin-offs are companies which have been created in order to commercialise IP arising out of a research institution where IP is licensed, involving a patent or copyright, from the research institution to the new firm to form the founding IP of the firm and staff may be seconded or transferred full or part-time from the research institution to the new firm" (p. 5-6). "Start-ups or indirect spin-off companies are companies set up by former or present university staff and/or former students drawing on their experience acquired during their time at the university, but which have no formal IP licensing or similar relationships to the university" (p. 6).
Shane	2004	"... a new company founded to exploit a piece of intellectual property created in an academic institution" (2004, p. 4).
De Coster and Butler	2005	"University spin-off companies are high-technology ventures that originate from research work in a university, resulting in the generation of intellectual property and, usually, subsequent involvement of key researchers" (p. 535).
Leitch and Harrison	2005	"Spin-outs are defined as new companies formed around a core technology discovered in a lab. The parent organisation sells, licences or somehow transfers the technology to the spin-out, which is often founded by researchers from the parent company or campus" (p. 259).

Many scholars have referred to the definition introduced by Smilor, Gibson, and Dietrich (1990, p. 63). "A spin-off is a company that is founded (1) by a faculty member, staff member, or student who left the university to start a company or who started the company while still affiliated with the university; and/or (2) around a technology or technology-based idea developed within the university".

The main differences and similarities concern the individuals who are involved as entrepreneurs, their relationships to the parent institution, and the knowledge or technology on which the business is based (Pirnay, Surlemont, and Nlemvo, 2003). Regarding the individuals, some studies have taken only faculty members or former employees such as researchers or lecturers into consideration (e.g., Carayannis, Rogers et al., 1998; Steffensen, Rogers, and Speakman, 2000), whereas other authors have also considered students or graduates as potential founders of an academic spin-off (e.g., McQueen and Wallmark, 1982; Smilor, Gibson, and Dietrich, 1990; Rappert, Webster, and Charles, 1999).

Many authors consider technology a key component of a spin-off, that is, technical ideas (McQueen and Wallmark, 1982), technology or technology-based ideas (Smilor, Gibson, and Dietrich, 1990), core technology (Carayannis, Rogers et al., 1998; Steffensen, Rogers, and Speakman, 2000; Nicolaou and Birley, 2003; Clarysse and Moray, 2004; Leitch and Harrison, 2005), or high technology (De Coster and Butler, 2005). The definitions given in previous studies, however, do not agree on the nature of the technology or the knowledge that must be transferred from the parent institution to the new spin-off company. It can be a core technology protected by patents or knowledge generated within a research project. Some scholars emphasise the technology transfer process (e.g., Shane, 2004). Furthermore, some studies address the role the parent institution plays in the spin-off process. Clarysse, Heirman, and Degroof (2000) assess the parent institution as playing an active role in initiating the spin-off.

2.2   Typologies of academic spin-offs

The definitions presented in the previous chapter do not have clear boundaries. In particular, if the entrepreneurial team can consist completely of students, for instance, any type of knowledge transfer is sufficient to meet the given definitions. Therefore, in the next paragraph, spin-off companies with different characteristics are framed. Several classifications and typologies have been developed to systematically describe the spin-off phenomenon (e.g., Stankiewicz, 1994; Pirnay, Surlemont, and Nlemvo, 2003; Hindle and Yencken, 2004; Wright, Clarysse et al., 2007). A typology generally outlines different conceptual types of organisations that may or may not exist, whereas a classification scheme categorises a set of phenomena into different subsets (Doty and Glick, 1994). The observed objectives often do not fully belong to any one ideal type; rather, this perspective allows for a more theoretical view of the phenomena and relationships. Although spin-off companies are often assumed to belong to high-tech ventures, they can also be established within a non-technology context (De Cleyn and Braet, 2009).

Pirnay, Surlemont, and Nlemvo (2003) proposed a typology based on two key discriminatory factors. These factors are (1) the academic status of the individuals who are involved in the new venture and (2) the type of knowledge that is transferred. Concerning academic status, a differentiation is made between entrepreneurs with a scientific background, such as professors or researchers, and founders without substantial research experience, more similar to students. Spin-offs with a scientific background typically try to commercialise research findings with promising prospects for future success. Therefore, researchers and occasionally students are engaged in this type of spin-off. With regard to types of knowledge, both codified knowledge and tacit knowledge can be transferred to the spin-off company. Codified knowledge is usually distinct from the researcher who developed it. Because it relies on formal information such as publications, research reports, or even patents, it is easily accessible and can be used or reproduced by other individuals. Tacit knowledge, such as experiences or technical expertise, is generally related to the researcher who acquired this type of knowledge within his or her academic career. Spin-off companies exploiting codified knowledge are primarily product-oriented, whereas ventures are based on tacit knowledge and often operate in the service sector (Nlemvo, Pirnay, and Surlemont, 2002).

Referring to Heirman and Clarysse (2004), Wright, Clarysse et al. (2007) distinguished between the VC-backed, prospector, and lifestyle types. According to this typology, the VC-backed spin-off is based on codified knowledge and, thus, is often protected by intellectual property. The prospector and the lifestyle types do not necessarily rely on patents or license agreements with the parent institutions. Although most visible and recognised, the VC-backed type comprises only a minority of spin-offs. Built on disruptive technology, it attracts investors to achieve growth and further development. Most spin-offs belong to the prospector or lifestyle category. Both types of companies start with a small entity and try to reach the break-even point quickly. The lifestyle type usually does not rely on a formal transfer of technology and is driven by market acceptance rather than investor acceptance, while the prospector type can be built on technology and attract a limited amount of external capital. The configuration of the spin-off company may have an impact on strategy and performance. Ventures based on codified knowledge might be more technology-driven, whereas spin-offs built on tacit knowledge are more market-driven (De Cleyn and Braet, 2009). For this reason, this study will consider basic characteristics of the ventures as possible influencing factors for strategy and performance.

Figure 2.1: Typology of university spin-offs

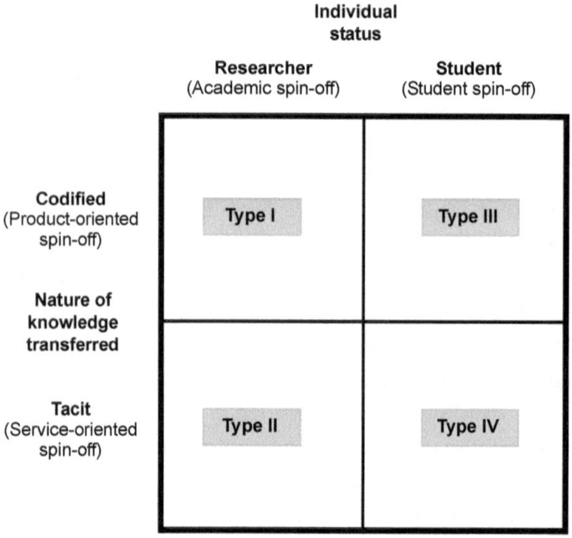

Source: Pirnay, Surlemont, and Nlemvo (2003, p. 361)

2.3 Definition of the target group

The target group of this study is research-based spin-off companies from academic organisations. To enable comparability with other studies, the main concepts of an academic organisation and a research-based spin-off company must be clearly specified. Therefore, within this study, a spin-off is defined as follows. "Research-based spin-offs are new ventures created to exploit commercially knowledge, technology, or research results developed within a research institution". This definition is based on Pirnay, Surlemont, and Nlemvo (2003), Klofsten and Jones-Evans (2000), and De Cleyn and Braet (2009). In addition to the definition, further analysis only considers companies that meet the following criteria:

a) New company: The spin-off must have the legal structure of a new firm. Neither establishing an extension or a subsidiary of the research institute nor transforming an existing company is considered a spin-off.

b) Parent organisation: The new company must have emerged from an academic organisation that is specified as a research institute financed mainly by public funding. This definition predominantly includes universities but also covers universities for applied science and colleges as long as they are engaged in research activities. Furthermore, the study considers non-university research organisations, such as the Max-Planck Gesellschaft in Germany. The given delineation does not comprise companies or private laboratories.

c) Knowledge and technology transfer: The spin-off commercialises scientific knowledge that may include technological innovations or patents and individual know-how. The knowledge was generated within an academic career in the research institute and transferred to the new company. The transfer can take place in several ways such as researchers joining the spin-off, transfer of protected intellectual property via sale or licensing agreements.

d) Profit-making objective: The founding of the new business should have generating profit as a main objective. Therefore, non-profit organisations are not considered.

2.4   Scope of prior studies

Companies that have emerged from the academic environment have recently attracted growing interest among scholars, as documented by several reviews of the literature (e.g., O´Shea, Allen et al., 2004; Rothaermel, Agung, and Jiang, 2007; Djokovic and Souitaris, 2008; Yusof and Jain, 2010). The majority of studies have been conducted in the US context, where commercialising new technology via spin-off companies has been a common path in the technology transfer process for much longer than in Europe (Clarysse, Wright et al., 2005). Most studies have only considered a single country or a particular research institute.

Different perspectives and frameworks have been applied and are identified by various terms such as university entrepreneurship, academic entrepreneurship, or university-level entrepreneurship. Previous research has been dominated by studies that focus on issues such as commercialisation strategies and technology transfer processes in research organisations, governmental support schemes, and legislative frameworks. These studies have been published in journals such as Management Science, The Journal of Technology Transfer, Research Policy, and The Journal of Business Venturing. For this reason, the majority of studies in the context of academic spin-offs have looked at factors that influence spin-off activity and the creation of new ventures.

In a comprehensive literature review, Rothaermel, Agung, and Jiang (2007) classified studies in the field of university entrepreneurship into four categories: (1) entrepreneurial research universities, (2) productivity of technology transfer offices, (3) new firm creation, and (4) environmental context. The first line of research, which represents approximately half of the field, looks at activities and schemes that encourage universities to become entrepreneurial. The second category deals with the impact of technology transfer offices as a gateway between the university and industry and accounts for approximately 10 per cent of the studies. The last category addresses environmental factors and accounts for 15 per cent of the research. The present study is related primarily to the third stream, that of new firm creation, which comprises approximately 25 per cent of the prior research in the area of university entrepreneurship. In this context, Rothaermel, Agung, and Jiang (2007) identified different measurements used in the examined studies that cover the number of new ventures established, their performance and their development as measured by indicators such as survival, growth, VC-funding, IPO, and specific characteristics of the companies. By focussing on new firm creation as a dependent variable, several possible influencing factors were identified, such as the university system, including faculty and technology transfer offices, the underlying technological base, investors, and the founding team and their network. With respect to the founding team, experience, qualification, social capital, team composition, and diversity are all relevant factors.

The methodological approaches that were applied most often include quantitative regression methods to measure the performance of technology transfer offices or research institutes in terms of spin-off formation rates. However, concerning the team level, quantitative empirical studies are rare as the majority of studies have conducted qualitative methods based on case studies (Rothaermel, Agung, and Jiang, 2007).

2.5 Theoretical perspectives

Some authors have noted that many empirical studies lack a well-grounded theoretical explanation for the observed relationships and phenomena (e.g., Nicolaou and Birley 2003; O'Shea, Allen et al. 2004; Rothaermel, Agung, and Jiang, 2007; Djokovic and Souitaris 2008). However, academic entrepreneurship is still in an early stage as an independent field of research where "a 25-year history is considered a very short time when compared with, for example, the 50-year history of strategy or the more than 225-year history of economics" (Rothaermel, Agung, and Jiang, p. 9). Furthermore, Davidsson and Wiklund (2001) noted that scholars in the entrepreneurship field often take theories from other research disciplines, as there is currently no consensus about an overarching theory of entrepreneurship.

The following paragraph outlines different theoretical views that have been applied in previous studies on research-based spin-off companies. The literature on academic spin-offs has acknowledged the high degree of heterogeneity and diversity among the examined companies and has noted that scholars used different theoretical lenses in their studies. Mustar, Renault et al. (2006) consider research-based spin-off companies as a subgroup of new high-technology ventures that face similar constraints and challenges when developing their businesses. Based on a review of the studies published since 1990 in the fields of entrepreneurship, strategy, innovation, and public policy, it is worth noting that they frame the research in a multi-dimensional taxonomy and identify three main theoretical perspectives.

A first group of studies examined spin-offs from an institutional perspective, analysing the relationship between the spin-offs and their parent organisations. In this context, researchers question how the mother institutes may affect the initial start-up composition, business model, and further development of the companies (e.g., Carayannis, Rogers et al., 1998; Steffensen, Rogers, and Speakman, 2000; Link and Scott, 2004; Lindelof and Lofsten, 2004; Wright, Birley, and Mosey, 2004; Clarysse, Wright et al., 2005).

Secondly, the framework outlined by Mustar et al. (2006) includes studies that adhere to a business model perspective. A subgroup examined the activities and portfolios of the companies and distinguished between predominantly service-oriented and product-oriented firms on one hand and market-oriented or technology-oriented strategies on the other hand (e.g., Chiesa and Piccaluga, 2000; Bower, 2003; Pirnay, Surlemont, and Nlemvo, 2003). The next subgroup focused on different forms of value creation through technology or knowledge transfer (e.g., Druilhe and Garnsey, 2004; Heirman and Clarysse, 2004). Another subset of studies considered the business models that were interested in the growth orientation of the ventures and differentiated between slow, fast, and transitional growers (e.g., Autio and Lumme, 1998).

Finally, the third view can be considered the resource-based perspective as the studies examine companies that have different resource configurations that influence their performance and competitive advantage. This group of studies refers, to some extent, to the resource-based view of the firm (Barney, 1991). The analysis of predictors of new venture performance has become an important aspect of entrepreneurship research. A number of studies have noted that initial resource stocks may have subsequent influence on organisational outcomes, including performance. Varying initial resources lead to varying outcomes at the organisational level. In this context, extant research has discussed tangible assets, such as financial capital or physical properties, as well as intangible resources, such as knowledge, routines, or collaboration activities (e.g., Radosevich, 1995; Westhead and Storey, 1995; Franklin, Wright, and Lockett, 2001; Shane and Stuart, 2002; Hindle and Yencken, 2004; Wright, Vohora, and Lockett, 2004) .

2.6 Performance in research-based spin-offs

Research at the firm level that investigates and analyses influencing factors for success and performance of new ventures has become fundamental in entrepreneurship research. In this context, it is important to assess those specific outcomes that may indicate successful companies. Whether the focus of the company is more on profitability, sustainability, or some other aspect depends on the founders', the owners', and/or the stakeholders' viewpoints. However, in previous research, a variety of measures and scales have been established and used that indicate a dependency on the developmental stage of the observed companies.

Prior studies have measured success by looking at the operating figures that reflect the performance, such as profitability, which can be measured by return on investment or by net profit (e.g., Roberts, 1991a; Hemer, Berteit et al., 2005). Another important aspect is the growth of the company, which is reflected by turnover or the number of employees (e.g., Ensley and Hmieleski, 2005; Walter, Auer, and Ritter, 2006). In this respect, research on new ventures must face the problem that these figures are often not available. In an early life-cycle phase, revenue streams might be low and the break-even point has very likely not been reached; thus, firms do not make any substantial profits. Growth rates might be high, as firms tend to be small at the beginning and grow rapidly relative to their size in the early years. Furthermore, data about these companies are not available in public databases. Therefore, self-reported or self-perceived measurements are often used; however, these data need to be considered in terms of objectivity, subjectivity, and reliability.[2] Spin-off success has also been measured by specific milestones, such as raising venture capital funding (e.g., Lockett, Murray, and Wright, 2002; Shane and Stuart, 2002; Zucker, Darby, and Armstrong, 2002), realising an initial public offering (e.g., Goldfarb and Henrekson, 2003; Shane, 2004) and external evaluations such as credit ratings (e.g., Egeln, Gottschalk et al., 2003). Finally, the survival of a company has often been used to assess spin-off success (e.g., Shane and Stuart, 2002; Shane, 2004).

Concerning the performance level, studies indicate that research-based spin-off companies and other new technology-based ventures or traditional start-ups differ. In the United States and Europe, academic spin-offs record lower failure rates or higher survival rates than do other new ventures, on average (e.g., Degroof and Roberts, 2004; Djokovic and Soutaris, 2008). Studies also reported that spin-offs are more likely to survive than other high-tech start-ups in the same industry (e.g., Mustar, 1997; Egeln, Gottschalk et al., 2002). The reasons for these trends, however, are not fully understood. The support mechanisms provided by the parent institution and governmental promotion schemes might yield an advantage during the start-up years. With regard to growth, the findings are inconsistent. For example, Ensley and Hmieleski (2005) reported that spin-offs have a higher annual growth rate in turnover than other technology-based ventures, and according to Egeln, Gottschalk et al. (2002), spin-offs grow faster in terms of employment compared to other firms belonging to the same industry sector. While a small percentage of spin-offs belong to the group of rapidly growing ventures, the majority tend to grow slowly (Autio, 1994; Stankiewicz, 1994; Mustar, 1997; Harrison and Leitch, 2005). Finally, in terms of productivity and externally measured credibility, some

---

[2] See Chandler (1993) for more details regarding the validity of subjective performance measures

researchers claim that spin-offs perform worse than traditional start-ups (e.g., Berndts and Harmsen, 1985; Egeln, Gottschalk et al., 2002), but according to Colombo and Delmastro (2002), the differences between academic spin-offs and traditional start-ups are negligible.[3]

The discussion regarding the success factors of spin-offs and the possible performance measurements leads to the question of which influencing factors might affect different outcomes at the organisational level. In this context, several frameworks have been introduced (e.g., Szyperski and Klandt, 1981; Kulicke, 1987; Unterkofler, 1989; Roberts, 1991a; Ndonzuau, Pirnay and Surlemont, 2002; Helm and Mauroner, 2007). In a literature review of the success factors of academic spin-off companies, Helm and Mauroner (2007) built on the framework for new venture creation introduced by Gartner (1985), noting three key categories. First, factors regarding the founders and the founding team were mentioned (individual/team level). This category includes demographic characteristics such as age and gender, which were not found to have a significant impact (Roberts, 1991a). However, skills, knowledge, and capabilities, also called human capital, significantly affect the performance of spin-offs, whereas personality and entrepreneurial motivation only partially influence success. The second category covers the organisational level, including general attributes of the spin-off itself, such as age or size. Furthermore, this aspect refers to the resource-based view stating that technological, financial, and social resources of the company have a significant impact on the performance (Helm and Mauroner, 2007). Finally, the discussion about factors regarding the environmental level incorporate the location of the company, institutional environment, market, and industry-specific aspects as well as the characteristics of and support from the parent institution (Helm and Mauroner, 2007).

This study focuses on human capital, mainly on the impact of the founding, entrepreneurial, and management team on organisational outcomes, such as firm performance, which will be discussed in the remainder of this chapter.

2.7 Boundaries of the entrepreneurial and management team

Previous research has, for the most part, agreed about the relevance of entrepreneurs, owners, and managers as the key resource assets and influence factors for the growth and performance of small companies and new ventures, in particular (e.g., Bruno, Leidecker, and Harder, 1987; Watson, 1995; Shane, 2000). In this context, studies have analysed the characteristics of founders and owner-managers with the entrepreneur as the unit of analysis (Davidsson and Wiklund, 2001). According to the literature, which has examined the human resources of new ventures, primary competencies of a high-technology company depend on the competencies

---

[3] See Ensley and Hmieleski (2005) for more details with regard to differences between spin-offs and regular start-ups

of the entrepreneurial team (Cooper and Bruno, 1977). Therefore, the growth and performance of spin-offs are closely linked to the characteristics of their founders and owner-managers.

High-tech companies and research-based spin-offs in particular are founded by a team rather than by a single individual (Roberts, 1991a), and high-growth companies are much more likely to be run by an entrepreneurial team as well (Kamm, Shuman et al., 1990). Literature in the field of venture capital, for example, posits that the entrepreneurial team of a company is the most influential factor with respect to decisions about investments (Macmillan, Siegel, and Narasimha, 1985; Hsu, 2007). In this sense, the entrepreneurial team is an intangible asset of the company (Cooper and Daily, 1997). Nevertheless, initially, entrepreneurship research studies regarding entrepreneurs and their attributes were most often conducted from an individual point of view. Recently, however, the team-level perspective has become increasingly important (Birley and Stockley, 2000; Chowdhury, 2005; Beckman, Burton, and O'Reilly, 2007). Although entrepreneurial teams have a significant impact on firm performance (Virany and Tushman, 1986; Kamm, Shuman et al. 1990; Eisenhardt and Schoonhoven 1990; Shane, 2004), they have not been extensively studied (Birley and Stockley, 2000), especially in the context of academic entrepreneurship.

Before discussing the literature and empirical findings on entrepreneurial and management teams in research-based spin-off companies, the boundaries of an entrepreneurial team must be clarified. Table 2.2 presents several existing definitions.[4] According to the definitions, the important aspects in determining an entrepreneurial team include being a member of the founding team, holding an equity stake, participating in decision making, and managing or running the firm on a full-time basis. The question is to which extent these definitions are applicable to research-based spin-offs.

Establishing an academic spin-off can take a long period of time that usually begins with an extensive research phase even before the firm is officially incorporated (Vohora, Wright, and Lockett, 2004). Furthermore, entrepreneurial teams in research-based spin-offs have a dynamic perspective. Vanaelst, Clarysse et al. (2006) distinguished between pre- and post-start-up teams. In addition, entrepreneurial teams in academic spin-offs usually evolve and change in composition over a period of time (Vanaelst, Clarysse et al., 2006). Thus, it is very likely that new members who join the initial team bring in new skills and competencies or replace the academic entrepreneurs who return to the parent institution. For this reason, there might be different entry modes for new members of the entrepreneurial team. A new team member can first be involved as a member of the management team and become a stakeholder of the company later. In this respect, it is very difficult to delineate the boundaries of a spin-off's entrepreneurial team at any specific point in time.

---

[4] Reviews on new venture team literature: Cooper and Daily (1997), Birley and Stockley (2000)

However, a research-based spin-off that is based on cutting-edge technology and groundbreaking scientific results very often has several stakeholders. The scientists who have been involved in the research projects may elect to join the new venture. Additionally, the representatives of the technology transfer office of the parent institution are very likely involved, for example, if patents or license agreements are concerned. More than likely, external organisations, such as venture capital funds, are represented on the board in the event of external funding. Based on the previous discussion, this study focuses on the executives of spin-off companies who are managing the business and responsible for the decision making. Thus, the terms entrepreneurial, management, and executive team are used synonymously. Further aspects, such as ownership and member of the founding team, will be considered separately. The empirical analysis does not specifically focus on teams with two or more individuals, spin-offs managed by only one executive are included as well.

Table 2.2: Definitions of an entrepreneurial team[5]

Authors	Year	Definition
Cooper and Bruno	1977	"…two or more full-time founders" (p. 20).
Kamm, Shuman et al.	1990	"…two or more individuals who jointly establish a firm in which they have a financial interest" (p. 7).
Eisenhardt and Schoonhoven	1990	"The founding top management team was defined as those individuals who were founders of the firm and who worked full time for the firm in executive-level positions at the time of founding" (p. 515).
Watson, Ponthieu, and Critelli	1995	"A venture team is two or more individuals who jointly establish and actively participate in a business in which they have an equity (financial) interest" (p. 394).
Ensley, Pearson, and Amason	2002	"…define the members of the top management team as being those individuals who met at least two of three conditions. They either were founders, currently held an equity stake of at least 10%, or were identified in some way as being actively involved in strategic decision making" (p. 372).
Chowdhury	2005	"An entrepreneurial team is often characterized as two or more individuals with equity interest jointly launching and actively participating in a business" (p. 730).
Harper	2008	"…a group of entrepreneurs with a common goal that can only be achieved by appropriate combinations of individual entrepreneurial actions" (p. 5).

---

[5] Further references: Gartner, Shaver et al. (1994), Vanaelst, Clarysse et al. (2006)

2.8 Previous findings about executives in research-based spin-offs

The entrepreneurship literature has discussed several facets of entrepreneurial and management teams. These facets include (1) general demographic characteristics, such as age and gender; (2) human-capital-related aspects of the team, including experience, education, and specific knowledge and capabilities; (3) team composition with respect to heterogeneity or homogeneity; and (4) evolution and dynamics with aspects such as entries and exits, tenure and turnover. In the next section, findings from previous research about entrepreneurial teams in research-based spin-off companies are discussed, including aspects regarding founder and ownership and specific characteristics of the executive teams that must be considered for the upcoming analysis.

Although executive teams of spin-off companies differ in terms of demographic attributes when compared with other start-up companies, the impact on performance is similar (Helm and Mauroner, 2007). Women are generally underrepresented and account just for 6 to 15 per cent of the founders, and only 25 per cent of the spin-off ventures include women (Kriegesmann, 2000; Egeln, Gottschalk et al., 2002). Nevertheless, previous studies do not report any empirical evidence for a significant positive or negative relationship between gender and spin-off success (Helm and Mauroner, 2007). Regarding the age of spin-off founders or the entrepreneurial team, prior research seems to agree that, again, there is no connection between age and the company's success (Roberts, 1991a; Kriegesmann, 2000). In fact, scholars have noted differences in terms of the average age between spin-off and other start-up founders but do not agree about the direction of the impact of these differences. Some studies report a lower average age for spin-off entrepreneurs (e.g., Brüderl, Preisendörfer, and Ziegler, 1996; Kay, May-Strobl, and Maaß, 2001), whereas others argue that the university system generates older entrepreneurs (e.g., Audretsch, 2000). Other demographical factors, such as religion, may influence the willingness to start a company but do not appear to affect the outcome (Kassicieh, Radosevich et al., 1997; Kriegesmann, 2000).

Substantial empirical evidence suggests that the characteristics of the entrepreneurial team, also called human capital, have an impact on spin-off success (Vohora, Wright, and Lockett, 2004; Helm and Mauroner, 2007). Nevertheless, the findings are contradictory. In the early years, the majority of spin-off entrepreneurs have a technical background (Vanaelst, Clarysse et al., 2006; Helm and Mauroner, 2007), and the entrepreneurial team often lack managerial and entrepreneurial experience (Berndts and Harmsen, 1985). Existing industry contacts or experience in the same industry sector within the team has been found to have a positive impact on spin-off success (Steinkühler, 1994; Shane and Stuart, 2002; Nicolaou and Birley, 2003; Scholten, 2006; Walter, Auer, and Ritter, 2006). However, Scholten (2006) reported a negative relationship between previous management experience and growth during the early years of a company. According to Roberts (1991a), entrepreneurs from research organisations who have prior commercial experience are also less successful, though studies regarding previous entrepreneurial experience have reached ambiguous conclusions. For example, Scholten (2006) noted a positive impact, while other studies did not support these findings and could not identify a significant relationship (e.g., Kassicieh, Radosevich et al., 1997; Kriegesmann, 2000; Shane and Stuart, 2002). Again, while some studies mentioned that educational level has a positive impact on performance (e.g., Hunsdiek, 1987; Steinkühler, 1994; Egeln, Gottschalk et al., 2003), Roberts (1991a) reported a U-shaped relationship, pointing out that additional qualifications beyond the masters level contributes negatively to firm performance.

The composition of an executive team of a research-based spin-off company has several specific characteristics that must be considered. Ensley and Hmieleski (2005) found that management teams of university-based start-ups tend to be more homogeneous regarding their educational background and functional and industry experience than those of other start-ups. Colombo and Delmastro (2002) could not identify significant differences between traditional start-ups and academic spin-offs. In many cases, the academic inventors establish the new ventures themselves. According to Roberts (1991a), founders can work partly for the new venture while still employed at the parent institution and shift their attention to the spin-off. However, Doutriaux (1987) and Roberts (1991a) expressed doubts about the effectiveness of this approach.

According to Vanaelst, Clarysse et al. (2006), entrepreneurial teams in academic spin-offs evolve over time, changing in their composition, so that external entrepreneurs and managers may partly or even completely replace the original founding team. In this context, the configuration of the founding team and the composition of the management team in a later development stage is discussed. Nicolaou and Birley (2003) identified three categories based on the roles of the researchers from the parent institution: (1) technological, (2) hybrid, and (3) orthodox spin-offs. In the first, technology is transferred to the new venture, but the researchers remain in the mother organisation without having a position in the spin-off. The second category defines a setting where the academician is involved in both the new firm and the research institute. In the third model, the researchers leave the parent organisation during the spin-off process. In this context, Olofsson and Wahlbin (1984) noted that rapidly growing spin-offs usually incorporate academics who have left the university environment. In this context, Van Dierdonck and Debackere (1988) differentiated between extra- and intrapreneurial spin-offs, which include researchers who leave or stay, respectively, in the parent institution.

Regarding the role and the background of entrepreneurs and managers in academic spin-offs, Radosevich (1995) outlined two distinct types: (1) the academic entrepreneur, also called the inventor or technology entrepreneur, and (2) the surrogate entrepreneur. An academic entrepreneur or inventor/technology entrepreneur is an individual who has participated in developing the technology on which the spin-off is based. Although his primary role was that of a lecturer or researcher, he now assumes the role of the entrepreneur in the new venture (Franklin, Wright, and Lockett, 2001). Surrogate entrepreneurship involves an independent and an external entrepreneur, which could be either an individual or an organisation from outside the research institute that initiated the company. The originator of the technology may keep his position in the research institute, in which case both the academic and surrogate entrepreneurs are simultaneously involved.

There is an ongoing debate regarding the effects of the performances of academic and external entrepreneurs and the dependency of their relationships on the development stage of the company or the environmental situation. Several studies examine surrogate entrepreneurship from a theoretical approach, which refers, basically, to resource endowments. Radosevich (1995) noted that surrogate entrepreneurs may play an important role in transferring technology. He argued that inadequate incentives and support mechanisms for scientists in research institutes may change the focus of the scientist's activities. Previous entrepreneurial experience and accumulated business knowledge, including professional networks, risk capital access, and strategic alliances, are advantages of the surrogate entrepreneur model.

Corresponding to a survey of 57 universities in the UK, Franklin, Wright, and Lockett (2001) determined that surrogate entrepreneurs could be an important source of commercial knowledge and skills, perhaps increasing the chances of success. On the contrary, based on a longitudinal case study of Clarysse and Moray (2004), external entrepreneurs from outside the research organisation may not understand the technology and, thus, may not be capable of successfully developing the company. Additionally, researchers rarely accept external managers on the team. Therefore, it is recommended that the original academic founding team include a coach and fill potential knowledge gaps in the learning process. This perspective is supported by Roberts and Hauptman (1986), who stated that academics in the new venture enable a more efficient technology transfer process.

Shane and Stuart (2002), in an analysis of 134 spin-off companies from MIT (US), found that new ventures with founders who have direct and indirect relationships with venture investors are most likely to receive venture funding. Additionally, these companies are less likely to fail. Furthermore, the authors argued that receiving funding is one of the most influential determinants of the likelihood of an IPO. For this reason, social capital assets that might be added by external entrepreneurs have a significant impact on fundraising. Wright, Vohora, and Lockett (2004), using case studies of four spin-offs in the UK, examined joint ventures with a company from the industry as a type of external partner. They concluded that creating a spin-off as a joint venture could help overcome resource constraints. External managers and entrepreneurs may contribute their business and commercial experiences as well as entrepreneurial knowledge, thus sending an important signal to external investors. Franklin, Wright, and Lockett (2001) suggested that a mixture of academic and surrogate entrepreneurs are an effective way of including both the technical know-how of the inventor as well as a market-oriented background.

2.9   Conclusion and research gap

Research-based spin-off companies differ in many respects from common start-ups (Shane and Stuart, 2002). The university environment implies a few peculiarities and constraints for establishing sustainable companies (Vohora, Wright, and Lockett, 2004). Academic spin-offs have similar characteristics as new technology-based ventures (Mustar, Renault et al., 2006), which are often challenged by several liabilities, such as newness (Stinchcombe, 1965; Freeman, Carroll, and Hannan, 1983), size (Aldrich and Auster, 1986; Brüderl and Schüssler, 1990), adolescence (Brüderl and Schüssler, 1990), uncertainty (Schröder, 2008), and owner dependence (Shane and Stuart, 2002). In addition, spin-off companies often rely on highly sophisticated research results and cutting-edge technology. Thus, they are usually established in a very early stage and need to continue developing their technologies and portfolios (Shane, 2004).

According to the introduced definitions, spin-off companies have specific features that differentiate them from new technology-based ventures. First, the transfer of technology or knowledge gathered within the research organisation implies that the mother institution is often involved in the founding process and might even be an additional stakeholder. Second, the transfer of people from the academic to the business environment implies certain characteristics of the entrepreneurial and management team. Former researchers often have a technical background and less commercial or business experience because they have worked in a non-commercial environment and focused on research and publication rather than transfer or commercialisation.

This study focuses on the second aspect. Many entrepreneurial teams lack the managerial and entrepreneurial skills to establish profitable ventures, as the scientists themselves are typically the founders of high technology companies emerging from academia. Ensley and Hmieleski (2005) found that management teams of university-based start-ups tend to be more homogeneous regarding their educational background and their functional and industry experience than other start-ups. This finding is especially true in the early years, when the majority of spin-off entrepreneurs and managers have a technical background (Helm and

Mauroner, 2007). Thus, the composition of the entrepreneurial teams are likely to change over time (Vanaelst, Clarysse et al., 2006) and involve experienced external entrepreneurs and managers who did not belong to the research institutions before joining the spin-offs (Franklin, Wright, and Lockett, 2001).

In this context, Radosevich (1995) identified two types of entrepreneurs and distinguished between the academic and the surrogate entrepreneur. Whereas academic entrepreneurs, also called inventor or technology entrepreneurs, actively participated in developing the technology, the external, or surrogate, entrepreneurs joined the spin-off from outside the research institute (Franklin, Wright, and Lockett, 2001). Most research has focused on the phenomenon of academic entrepreneurs, but only a few studies have studied the role and impact of external entrepreneurs within the executive teams (Mosey and Wright, 2007). There is substantial empirical evidence that the characteristics of the entrepreneurial team have an impact on spin-off success (e.g., Vohora, Wright and Lockett, 2004; Helm and Mauroner, 2007), but the findings are contradictory. For this reason, research must consider both the direct and indirect relationships between team characteristics and performance and examine how team characteristics may influence the internal processes of the organisations.

This need is the starting point of this study, which looks at different facets of academic entrepreneurship and addresses one specific aspect: the effect of executive teams on organisational outcomes. This study explores the effects of resource endowments on organisational outcomes but concentrates on the human capital aspect. In particular, emphasis is placed on the effect of the executives' previous entrepreneurial experiences, an area that is attracting more and more interest among entrepreneurship scholars (e.g., Chandler and Jansen, 1992; Dyke, Fischer, and Reuber, 1992; Westhead, Ucbasaran, and Wright, 2005). In this discussion, an essential differentiation is made between habitual entrepreneurs who have previous business-ownership experience and novice entrepreneurs who are running their business without any prior entrepreneurial experience (Ucbasaran, Alsos et al., 2008). There is a lack of research addressing the extent of habitual entrepreneurship in companies that emerged from academia (Mosey and Wright, 2007).

Furthermore, the findings from previous research concerning the relationship between prior entrepreneurial and business-ownership experience on firm performance are not consistent (e.g., Stuart and Abetti, 1990; Westhead and Wright, 1998). Habitual entrepreneurs do not, by default, perform better than novice entrepreneurs (Ucbasaran, Westhead, and Wright, 2006) simply because they might have been able to enlarge their knowledge and capabilities in prior businesses. They can also adopt routines and mindsets that limit the identification of opportunities and the development of the emerging business in a new and different environment (Starr and Bygrave, 1991). In view of this inconclusive or weak support for a positive impact of previous entrepreneurial experience on firm performance, the perspective regarding this debate must be broadened. Hambrick and Mason (1984) discuss the relationships between top management characteristics and the strategic decisions of the company as well as the organisational outcomes. Although this framework, called the upper echelons theory, was introduced more in the context of large corporations, applying it in an entrepreneurial setting may add an interesting perspective as it outlines a mediating effect of the characteristics of the management team on firm performance through the team's strategic choices.

In summary, most of the prior research on the entrepreneurial teams of academic spin-offs from research institutes has focused on the inventor as an entrepreneur. The effects of surrogate or habitual entrepreneurs on the development and performance of research-based spin-offs have been mainly analysed qualitatively; thus, further quantitative empirical studies are needed. The theoretical approach of this research stream, which studies performance and potential influence factors, refers primarily to the resource-based view (Barney, 1991). In this context, it must be mentioned that (1) the resource-based view is not really interested in processes, even though they might be effected by the characteristics of the management team and have a potential influence on organisational outcomes, and (2) the resource-based view does not assume that certain experiences, which are considered to be resources, could also be a liability for the company (Cannella, Finkelstein, and Hambrick, 2008). Therefore, the theoretical approach must be extended. The previous discussion leads to the following central research question, which guides the analysis of this study:

Research question:

How do executive teams in research-based spin-off companies influence organisational outcomes?

Because literature on executive teams in research-based spin-off companies is limited, studies with focus on new technology-based ventures and small and young companies were considered as well. In table 2.3, the literature that deals with different aspects of founding, entrepreneurial, and management teams is presented. The overview includes methods, measures, and key findings regarding the direct impact on performance and venture development. While several findings were discussed in this chapter, some references will be addressed in the following chapters.

Table 2.3: Overview of studies on entrepreneurial and management teams

No.	Authors	Topic	Sample	Data	Analysis	Context	Measures	Main findings
1	Amason, Shrader, and Tompson (2006)	Top management team characteristics and new venture performance	174 new ventures	Archival data (longitudinal)	Hierarchical regression	US	Sales growth, profitability, stock market returns	Novelty interacts with characteristics of top management team and affects performance. Interaction of novelty and diversity is positively associated with performance.
2	Aspelund, Berg-Utby, and Skjevdal (2005)	Initial internal resources as antecedents of survival	80 NTBV	Archival data, follow-up surveys (longitudinal data)	Linear regression	Norway, Sweden	Survival (hazard rate)	Smaller teams and higher degree of diversity increase likelihood of survival.
3	Baron and Ensley (2006)	Opportunity recognition of experienced and novice entrepreneurs	88 experienced, 106 novice entrepreneurs	Questionnaire	Content analysis	US	Business opportunity attributes	Business opportunity prototypes of experienced entrepreneurs are better defined and more related to factors regarding starting and running new venture.
4	Barringer, Jones, and Neubaum (2005)	Characteristics of rapid-growth firms and their founders	50 rapid-growth firms, 50 slow-growth firms	Case studies	Content analysis	US	Rapid growth, slow growth	Founder characteristics and firm attributes are important to generate rapid growth.
5	Bates (1990)	Owner human and financial capital inputs and business longevity	4,429 small firms	Archival data, survey	Multivariate regression, discriminant analysis	US	Survival, access to debt	Human capital inputs partially cause financial capital inputs. Owner education and financial capital inputs explain company longevity.
6	Batjargal (2007)	Social capital, human capital and performance	94 internet ventures	Telephone interviews	Logistic regression analysis	China	Survival	Interaction of social capital and start-up experience is negatively associated with firm performance.
7	Beckman, Burton, and O'Reilly (2007)	Top management team demographic characteristics and firm outcomes	161 young high-tech firms	Interviews, survey, archival data	Event history analysis, cox proportional hazards models	US	VC-funding, IPO	Functional and affiliation heterogeneity are related to positive outcomes. Adding knowledge and experience increases likelihood of VC and IPO.

Table 2.3: Overview of studies on entrepreneurial and management teams (continued)

No.	Authors	Topic	Sample	Data	Analysis	Context	Measures	Main findings
8	Boeker and Karichalil (2002)	Firm and individual characteristics and founder departure	78 young semiconductor firms	Archival data (longitudinal)	Hazard rate model, regression analysis	US	Founder departure	Founder departure is accelerated by firm size, decreases with founder ownership and board membership, and has a U-shaped relationship with firm growth.
9	Boeker and Wiltbank (2005)	Factors influencing changes in the top management of start-ups	86 semiconductor ventures	Archival data, interviews	Regression analysis	US	Changes in the top management team	Very low or very high firm growth increase changes in top management. VC increases and managerial ownership decreases changes.
10	Bruton and Rubanik (2002)	Resources of the firm and firm growth	45 high-tech firms	Structured interviews	Regression analysis	Russia	Growth (employment)	Larger teams are likely to generate greater resources leading to greater firm performance.
11	Chandler and Hanks (1994)	Moderating effects of founders' entrepreneurial and managerial competencies	155 manufacturing firms	Questionnaire	Moderated regression analysis	US	Perceived growth: market share, cash flow, sales	Individual competencies moderate relationships between opportunity quality, acces to resources, and firm performance.
12	Chandler and Hanks (1998)	Substitutability of founders' human and financial capital	102 firms	Questionnaire	Regression analysis	US	Sales/earnings, growth	Human and financial capital are substitutable. Firms with high founder human capital and low financial capital perform like firms with low human capital and great financial capital
13	Chandler and Jansen (1992)	Founders' entrepreneurial, managerial, and technical competences and performance	134 firms	Questionnaire	Factor analysis, discriminant analysis	US	Profitability, growth	Successful founders perceive themselves as competent in entrepreneurial, managerial, and technical-functional roles.
14	Chandler, Honig, and Wiklund (2005)	Initial team size, diversity, firm stage, environmental dynamism, and turnover in venture teams	408 new ventures, 124 established firms	Questionnaire	Binary logistic and moderated regression analysis	Sweden, US	Profitability, sales growth	Team size and diversity affects turnover. Departures are increasingly associated with sales growth, additions become increasingly negative.

Table 2.3: Overview of studies on entrepreneurial and management teams (continued)

No.	Authors	Topic	Sample	Data	Analysis	Context	Measures	Main findings
15	Clarysse and Moray (2004)	Spin-off development phases	1 academic spin-off	Single longitudinal case study	Content analysis	Belgium	Entrepreneurial team formation	Entrepreneurial team formation is a process of self-organized punctuated equilibrium.
16	Colombo and Grilli (2005)	Relation between growth and human capital of founders	506 NTBV	Archival data, questionnaire	OLS regression analysis	Italy	Growth (employment)	Education and prior work experience affect growth. Management education, scientific and technical background positively influence growth.
17	Colombo, Delmastro, and Grilli (2004)	Characteristics of founders and initial startup size	391 NTBV	Archival data, questionnaire	Linear regression analysis	Italy	Initial firm size (number of employees)	Founders' human capital has a positive impact on firms' initial size. Specific human capital has greater impact than general human capital.
18	Cooper, Gimeno-Gascon, and Woo (1994)	Initial human and financial capital and performance of new ventures	1,053 new ventures	Questionnaire (longitudinal data)	Discriminant analysis, logistic regression	US	Survival, growth	General human capital influences survival and growth. Management know-how has limited effect.
19	Cressy (1996)	Human capital, financial assets, and venture survival	2,000 start-ups	Archival data	Regression analysis	UK	Survival	Individual assets influence survival. Financial capital does not affect survival.
20	Daily and Dalton (1992)	Founder-managed, professional-managed ventures, and financial performance	186 small companies	Archival data, telephone interviews	Univariate/ multivariate regression analysis	US	Price/earnings ratio, return on assets, return on equity	Financial performance of founder-managed and professional-managed firms does not differ significantly.
21	Davidsson and Honig (2003)	Social and human capital and nascent entrepreneurial activities	380/608 individuals	Telephone survey, follow-up interviews (longitudinal data)	Rregression analysis	Sweden	Opportunity discovery/ exploitation, sales, profitability	Human and social capital increases likelihood of becoming nascent entrepreneur, small positive effect on sales and profitability.
22	Dubini (1989)	Entrepreneurial team characteristics and performance	151 ventures	Archival data	Cluster analysis, regression analysis	US	Sales volume, profitability	Different entrepreneurial team characteristics predict performance for different clusters.

Table 2.3: Overview of studies on entrepreneurial and management teams (continued)

No.	Authors	Topic	Sample	Data	Analysis	Context	Measures	Main findings
23	Eisenhardt and Schoonhoven (1990)	Founding team, strategy, environment, and growth	92 technical-based semiconductor ventures	Structured interviews, archival data (longitudinal data)	Multivariate regression analysis	US	Growth (sales)	Main and interaction effects for the founding top-management team and market stage on firm growth, no significant effect of technical innovation on growth.
24	Ensley and Hmieleski (2005)	Top management team composition, dynamics, and performance	102 high-tech university start-ups, 154 NTBV	Questionnaire	Discriminant analysis, multiple regressions	US	Team composition and dynamics, performance	Top management teams of university start-ups are more homogenous and perform lower than independent new ventures.
25	Ensley, Pearson, and Amason (2002)	Top-management interaction and new-venture performance	70 new ventures (Inc. 500)	Questionnaire	Structural equations modelling (LISREL), hierarchical regression	US	Sales growth, profitability	Affective conflict is negatively associated with performance, cohesion is negatively related to affective conflict. Cohesion among the management team is positively related to performance.
26	Ensley, Carland, and Carland (1998)	Entrepreneurial team skill heterogeneity, functional diversity, and new venture performance	88 fast growing firms	Questionnaire	Regression analysis	US	Revenues, sales growth, profitability	Team heterogeneity is negatively related to growth, diversity and functional background are negatively related to revenue.
27	Florin, Lubatkin, and Schulze (2003)	Influence of human and social capital on venture funding and performance	275 IPO firms	Archival data	Moderated regression analysis	US	Accumulated financial capital, sales growth, return on sales	Links between human resources and performance and between financial capital and resources vary with the level of social resources.
28	Gimeno, Folta et al. (1997)	Entrepreneurial human capital and performance and persistence	1,547 entrepreneurs of new businesses	Questionnaire	Censored regression, grouped data regression	US	Exit from business, economic performance	Specific human capital influences performance and survival, general human capital influences performance more than survival.
29	He (2008)	Performance of firms with founder-CEOs and professional CEOs	1,143 IPO firms	Archival data (longitudinal)	OLS regression analysis, fixed effects model	US	Return on assets ratio, survival	Founder-CEOs perform better than professional managers.

Table 2.3: Overview of studies on entrepreneurial and management teams (continued)

No.	Authors	Topic	Sample	Data	Analysis	Context	Measures	Main findings
30	Hitt, Biermant et al. (2001)	Direct and moderating effects of human capital on strategy and performance	93 professional services firms	Archival data, interviews	Regression analysis	US	Profitability	Human capital has curvilinear effect on performance. Human capital moderates links between strategy and firm performance.
31	Hsu (2007)	Prior start-up founding experience, organizational capital, and venture capital funding	149 early stage technology-based start-up firms	Survey	Multivariate regression analysis	US	Venture capital funding, venture valuation	Prior founding experience increases the likelihood of VC-funding and venture valuation. Fonders' ability to recruit executives is positively associated with venture valuation.
32	Knockaert, Ucbasaran et al. (2011)	Management team composition, knowledge transfer and performance	9 spin-offs	Case studies	Content analysis	Belgium		Tacit knowledge is most effectively transferred to the new company if researchers join the spin-off team.
33	Kulicke (1985)	Factors influencing growth phase	83 NTBV with 16 spin-offs	Survey	Descriptive analysis	Germany	Growth in sales and employees	Lack of professional experience inhibits growth.
34	Macmillan, Siegel, and Narasimha (1985)	Evaluation criteria for venture capitalists	100 venture capitalists	Questionnaire	Factor analysis, cluster analysis	US	Funding decision	Quality of the entrepreneur determines the funding decision.
35	Montgomery, Johnson, and Faisal (2005)	Human capital, financial capital, and business success	1,506 persons	Archival data	Regression analysis	US	Survival, revenues, earnings	Human capital increases the probability of self-employment, but not the likelihood of success.
36	Preisendorfer, and Voss (1990)	Entrepreneurial age, human capital and survival	78,441 small traders	Archival data	Bivariate analysis, multivariate analysis	West Germany	Survival	Concave relationship between founder age and organisational mortality. Human capital is more relevant in branches where industry experience is required.
37	Radosevich (1995)	Surrogate entrepreneurs in spin-offs	3 spin-offs	Case studies	Content analysis	US		Surrogate entrepreneurs can help to commercialise technology.

Table 2.3: Overview of studies on entrepreneurial and management teams (continued)

No.	Authors	Topic	Sample	Data	Analysis	Context	Measures	Main findings
38	Roberts (1991)	Success factors of spin-offs and NTBV	Several studies, e.g., 156 MIT spin-offs, 82 non-MIT spin-offs	Survey, case studies	Descriptive analysis	US	Sales growth, survival, profitability	Business experience of founders is positively related to early growth. Diversity of human capital is associated with performance.
39	Robinson and Sexton (1994)	Education, experience and self-employment success	21,352 self-employed people	Archival data	Probit regression analysis	US	Earnings potential	General education has a strong positive effect on self-employment success. Experience has a positive but small effect.
40	Roure and Keeley (1990)	Influence of management, firm's strategy and environment	36 NTBV	Archival data	Multivariate linear regression	US	Internal rate of return	Complete management team has positive influence on success.
41	Samsom and Gurdon (1993)	University scientists as entrepreneurs	22 high-tech ventures	Case studies	Descriptive analysis	US		Conflicts between business and scientific cultures constrain team work. Researchers consider lack of management skills as hurdle for success.
42	Shane and Stuart (2002)	Social capital of company founders	134 MIT-spin-offs	Archival data	Multivariate regression	US	VC financing, IPO, failure	Founders with relationships to venture investors are most likely to receive venture funding. VC-funding increases the likelihood of an IPO.
43	Shrader and Siegel (2007)	Human capital and firm performance	198 high-tech firms	Archival	Regression analysis	US	Profitability, sales growth	Strong link between team experience and strategy, fit between team experience and strategy determines long-term performance.
44	Steinkühler (1994)	Incubator centres and success of spin-offs	34 spin-offs, 35 NTBV		Bivariate, factor, and path analysis	Germany	Growth, productivity, profitability	Founders' industry experience and qualification is positively associated with growth.

Table 2.3: Overview of studies on entrepreneurial and management teams (continued)

No.	Authors	Topic	Sample	Data	Analysis	Context	Measures	Main findings
45	Stuart and Abetti (1990)	Entrepreneurial and management experience and early performance	52 NTBV	Questionnaire	Factor analysis, linear regression	US	Growth, profitability, productivity	Advanced education beyond bachelor is negatively, entrepreneurial experience is positively related to performance.
46	Ucbasaran, Westhead, and Wright (2009)	Entrepreneurs' prior business ownership experience and opportunity identification	630 entrepreneurs	Questionnaire	Probit analysis, binomial regression	UK	Opportunity identification, innovativeness	Business ownership experience positively affects identification of business opportunities and innovation.
47	Vanaelst, Clarysse et al. (2006)	Team heterogeneity in different phases of the spin-off process	10 academic spin-offs	Questionnaire, interviews, case studies	Content analysis	Netherlands	Team development (heterogeneity)	Teams evolve over time and change composition. Team's diversity changes in different stages of the spin-off process.
48	Virany and Tushman (1986)	Changes in top-management teams	59 technical based ventures	Archivial data, Interviews	Descriptive analysis	US	Return on assets	Executive succession is important for adopting to environmental challenges.
49	Vohora, Wright, and Lockett (2004)	Development of spin-offs	9 university spin-offs	Case studies	Content analysis	UK		General human capital positively influences spin-off performance.
50	Weinzimmer (1997)	Team heterogeneity, size, age of team members	74 firms		Correlation analysis		Sales growth	Functional heterogeneity, team size, and average age of team members are positively related to performance.
51	Wright, Vohora, and Lockett (2004)	Joint ventures and venture capital investors	4 spin-outs	Case studies	Descriptive analysis	UK	Business development	Joint venture spin-outs commercialise faster, more flexible, and less risky than venture backed university start-ups.
52	Zucker, Darby, and Armstrong (2002)	Publication and patents records and performance	112 "star" scientists	Survey	Multiple regression	US	Number of patents and citations	Number of academic publications is positively related to firm performance.

3 Theoretical framework

The previous chapter discussed contradictory findings regarding success factors in new technology-based ventures and research-based spin-offs in particular. In view of these results, this study broadens the presented view and takes process-related aspects in the company into consideration as well. The main focus is on aspects of human capital and the impact of the executive team on organisational outcomes. The following chapter outlines the theoretical background for further analysis and connects the entrepreneurship literature to the strategic management field. The structure is as follows. First, referring to the growing field of strategic entrepreneurship, important strategic processes for research-based spin-off companies are identified. In the next section, the areas of strategy content and strategic process research are distinguished and briefly introduced to position the examined topic in the field. The subsequent section narrows the scope of this study, presents literature on strategic decision making, and discusses the concept of rationality as the dominant logic in many studies. In the next step, the upper echelons framework is described and introduced as a theoretical framework as it helps to explain the relationships between the characteristics of the executive team, strategic decision making behaviour, and firm performance. Finally, the last section outlines the intended contributions of the underlying study.

3.1 Entrepreneurship and strategic management

Predicting performance is a key issue in the strategic management literature (Venkatraman and Ramanujam, 1986) and the entrepreneurship field (Shane and Venkataraman, 2000). A growing body of studies discusses complementary contributions of these two research streams (Ireland, Hitt, and Sirmon, 2003).[6] Some authors argue that these two fields should be more interrelated to better benefit from each other (e.g., Meyer and Heppard, 2000; McGrath and MacMillan, 2000). While the focus of interest in entrepreneurship research is on opportunities and the question of how to identify and exploit them, the strategic management field examines how companies can achieve a competitive advantage in their specific market environment. In this context, exploring, discovering, and developing new opportunities can also be a promising path for established companies to compete successfully (Zahra and Dess, 2001; Venkataraman and Sarasvathy, 2001; Kuratko, Ireland et al., 2005). While small firms and, in particular, new ventures are perceived as capable of exploring new opportunities, they

---

[6] See also Hitt, Ireland et al. (2001) and the special issue in Strategic Management Journal 2001 on Strategic Entrepreneurship

often lack the knowledge and skills to exploit them efficiently to sustain a competitive advantage. On the contrary, large corporations are said to be more efficient in exploiting rather than exploring opportunities. Therefore, the strategic entrepreneurship field postulates that large established corporations must consider entrepreneurial approaches in their strategic orientations, whereas young and small companies should start thinking strategically right from the beginning of their businesses (Ireland, Hitt, and Sirmon, 2003).

In the context of research-based based spin-off companies, opportunities play a central role as the portfolio is often based on cutting-edge technology, intellectual property, and/or new scientific results. This point of view makes clear that an entrepreneurial approach might be important for success, especially in the early stages when new products are being developed and introduced in the market. The question of how to exploit these opportunities efficiently seems to be a crucial aspect and indicates to the need to think strategically about commercialisation and management processes. The specific characteristics of the entrepreneurial team in a research-based spin-off company might be a hurdle for exploiting opportunities because, as outlined in the previous chapter, academic entrepreneurs often lack commercial knowledge and business experience. Therefore, this study is positioned at the intersection of entrepreneurship and strategic management and addresses different process-related questions. On one hand, how important is an entrepreneurial approach for research-based spin-offs and what are potential influencing factors? On the other hand, how important are comprehensive and intensive planning activities for academic spin-offs and what determines them? Acting entrepreneurial or planning strategically, do these two perspectives contradict or benefit each other? To this end, this study considers the antecedents and consequences of an entrepreneurial strategic posture and strategic planning processes within the organisations.

3.2 Strategic content and process research

Chandler (1962, p. 23) defines strategy as "...the determination of the basic long-term goals and objectives of an enterprise, and the adoption of courses of action and the allocation of resources necessary for carrying out these goals". According to this definition, strategy follows a formal planning process. Recent literature has already critically discussed the benefit of planning activities, such as business planning, especially in the context of start-ups or emerging ventures. For instance, Brinckmann, Grichnik, and Kapsa (2010, p. 24) formulated the provoking question, "Should entrepreneurs plan or just storm the castle?" Regarding conflicting empirical results in the field, they conducted a meta-analysis and conclude that planning is useful but depends on contextual factors, such as the age or newness of the firm. To enable a discussion about whether an entrepreneurial strategic posture or formal strategic planning activities are important for research-based spin-off companies, the following section develops the theoretical background for further analysis. Therefore, research on strategic decision making as a subset of strategy process research within the strategic management field is briefly introduced. The strategic management literature addresses two topics: the content of strategy and the process behind generating and implementing strategy (Rajagopalan, Rasheed, and Datta, 1993). The differentiation of strategy process and strategy content research goes back to the origins of strategic management research, such as Chandler's "Strategy and Structure" (1962), Ansoff's "Corporate Strategy" (1965), and Andrews' "The Concept of Strategic Management" (1971). In the early stages, research only analysed such aspects in large corporations (Analoui and Karami, 2003). Miles, Snow et al. (1978), and Porter (1980) fostered the development of strategy content research, which, since then, has been extensively examined both theoretically and empirically with respect to subjects such as mergers, acquisitions, diversification, and product-market combinations.

For diversified companies, various levels of strategy must be considered. On one hand, the corporate strategy as a type of overarching strategy looks at all business units of the firm and decides the composition of the business portfolio, the allocation of resources, and the structure of the company itself. In this context, Porter (1987, p. 43) states, "Corporate strategy concerns two different questions: What business the corporation should be in and how the corporate office should manage the array of business units. Corporate strategy is what makes the corporate whole add up to more than the sum of its business unit parts". On the other hand, strategy at the business level focuses on a single product or business unit to achieve a competitive advantage in the field. Campbell and Faulkner (2003, p. 12) note that "... finding a strategy that is better than that of your competitors, thus enables you to make repeatable profits from selling your products or services". For small companies, these two strategy levels are often equivalent and do not need to be distinguished. In most cases, research-based spin-offs begin as a small entity but might have the potential to grow quickly; thus, a differentiation might become relevant in a later stage. In addition, functional strategies as a third aspect include issues such as marketing, human resources, financial, legal, and/or information technology management, as they are more short-term in nature and are linked to corporate strategy.

Whereas research on strategy content analyses existing strategies from different perspectives, strategy process research examines the organisation and tries to explain how strategies evolve, which influencing factors exist, and how different factors interact with each other, such as the impact of the management team. Research on strategic processes has evolved more recently and has generated a multitude of different approaches.[7] In a comprehensive literature review on strategy process research, Hutzschenreuter and Kleindienst (2006) identified three key aspects: (1) antecedents, (2) processes, and (3) outcomes. The first category includes factors that influence strategic processes. In particular, this encompasses the environmental characteristics, such as uncertainty, complexity, munificence, dynamism, and the strategic position of the company; static organisational characteristics, such as size, age, and technology; and dynamic characteristics, such as routines, cultures, and values. Furthermore, past performance is considered an important antecedent. Concerning the second aspect, the framework divides strategy processes into three main parts comprising static and cognitive characteristics of the decision makers, such as age, experience, and origin. Issue-related aspects such as complexity and urgency, as well as process-related attributes such as comprehensiveness, are also mentioned. Finally, Hutzschenreuter and Kleindienst (2006) framed possible outcomes of strategy processes in the same way as the antecedent factors.

Strategy process research has either been prescriptive or descriptive (Hutzschenreuter and Kleindienst, 2006), meaning that it either describes how a strategy should be or how it is actually observed (Mintzberg and Lampel, 1999). Strategic process research has focused on main aspects such as planning methods and strategic decision making (Huff and Reger, 1987; Eisenhardt and Zbaracki, 1992). Another distinction in this context has been made between the generation and implementation of strategies in organisations (Andrews, 1971). Mintzberg and Lampel (1999) have clustered the existing literature on strategy process research into ten different schools of strategy. The papers in these categories emphasise different aspects of strategy formation. There are three prescriptive categories (design school, planning school, positioning school), six descriptive categories (entrepreneurial school, cognitive school, learning school, power school, cultural school, environmental school), and one type that combines both approaches (configuration school). Table 3.1 gives an overview of the different schools and outlines the key issues of each stream. The discussion between the most influential process schools – planning and learning – states from the perspective of the planning fraction that formal planning activities imply favourable effects irrespective of whether the environmental situation is changing or not (Ansoff, 1991, 1994). The learning school points out that organisational plans must be more flexible and must be adjusted, especially when dynamic changes in the environment occur (Mintzberg, 1991, 1994a). The entrepreneurial school, as a descriptive approach, is related to the work of influential scholars

---

[7] See Chandler (1962), Ansoff (1965), Bower (1970), Andrews (1971), Mintzberg (1978), Quinn (1980), Burgelman (1983), Miller (1993), Noda and Bower (1996), Iaquinto and Fredrickson (1997), Nutt (1998), Rindova (1999), Lovas and Ghoshal (2000), Golden and Zajac (2001), Farjoun (2002), Hiller and Hambrick (2005)

such as Schumpeter (1934) and focuses on the entrepreneur as a key factor in influencing the strategy formation process. Thus, intuition and vision, rather than detailed planning, play an important role (Mintzberg and Lampel, 1999; Kraus and Kauranen, 2009). The planning, learning, and entrepreneurial school are relevant for this study. Furthermore, some aspects refer to the environmental school of strategy that considers the environmental situations as a key influence factor for strategic processes.

Table 3.1: Mintzberg´s schools of strategy

Category	Key characteristics
Design School (prescriptive)	Strategy formation aims to achieve a fit between internal strengths and weaknesses as well as between external opportunities and threats. Origin: e.g., Selznick (1957)
Planning School (prescriptive)	Strategy is a formal planning process with defined steps and explicit methods. Origin: e.g., Ansoff (1965)
Positioning School (prescriptive)	Strategy development is an analytical process that looks at the industry situation and defines a strategic competitive position. Origins: e.g., Hatten and Schendel (1977), Porter (1980)
Entrepreneurial School (descriptive)	Strategy is based on the vision and perspective of the entrepreneur and depends on intuition rather than on formal plans. Origin: Schumpeter (1934)
Cognitive School (descriptive)	Strategy building is a mental process and based on individual perceptions and cognitive mindsets. Origins: e.g., March and Simon (1958)
Learning School (descriptive)	Strategy formation is an emergent learning process. Origin: e.g., Cyert and March (1963)
Power School (descriptive)	Strategy making is based on negotiation and power and can be applied within (micro) or outside (macro) the organisation. Origins: e.g., Allison (1971), Astley (1984)
Cultural School (descriptive)	Strategy formation is a social process and refers to culture. Origin: e.g., Rhenman (1973)
Environmental School (descriptive)	Strategy development depends considerably on the environmental situation as a key influencing factor. Origin: e.g., Hannan and Freeman (1977)
Configurational School (prescriptive and descriptive)	Strategy is a transformational process and can combine several aspects of different streams. Origin: e.g., Chandler (1962)

Sources: Mintzberg and Lampel (1999), Kraus and Kauranen (2009)

3.3   Strategic decision making

This study focuses on aspects of strategic decision making behaviour as a part of strategic process research, which is discussed in more detail in the following paragraph and includes aspects of strategic planning and the concept of entrepreneurial strategy making. According to Mintzberg, Raisinghani, and Theoret (1976, p. 246), a strategic decision can be characterised as "important, in terms of the actions taken, the resources committed, or the precedents set". This study examines how executives' characteristics on the team level affect strategic decision making behaviour at the firm level.

Different frameworks that try to structure the field of strategic decision making have been introduced (e.g., Allison, 1971; Mintzberg, 1973; Chaffee, 1985; Lyles and Thomas, 1988; Hart, 1992). Similar to the overview of strategic process literature introduced by Hutzschenreuter and Kleindienst (2006) and discussed in the previous paragraph, Rajagopalan, Rasheed, and Datta (1993) developed an integrative framework based on a review of studies that focus only on the strategic decision making aspect. They distinguish three categories of factors that influence the decision making process. As a first group, they identify environmental factors such as uncertainty, munificence, dynamism, and complexity (e.g., Dess and Beard, 1984; Sharfman and Dean, 1991). Second, they consider organisational attributes such as firm structure (e.g., Covin, Slevin, and Schultz, 1994), size (e.g., Yasai-Ardekani and Nystrom, 1996), and age (e.g., Forbes, 2005), as well as characteristics of the management team such as size (e.g., Iaquinto and Fredrickson, 1997) and heterogeneity (e.g., Ferrier, 2001). Other factors can include technological (e.g., Molloy and Schwenk, 1995) or the cultural aspects. Finally, factors specifying the decision itself belong to these antecedents. Figure 3.1 shows the entire framework, including the specific issues of each factor and the direct and indirect relationships that have been examined in the reviewed studies. The overview makes clear that the discussed factors can have a direct impact on processes and an indirect effect on economic outcomes. Rajagopalan, Rasheed, and Datta (1993) noted, however, that research about characteristics of the management team and their impact on strategic decision making is very limited, especially in the context of small firms and new technology-based ventures.

Based on the central assumption that human behaviour is driven by some type of intention and, thus, follows a rational logic, researchers coined the term *rational action* (e.g., March and Simon, 1958; Allison, 1971). This perspective assumes that human beings are goal-oriented and seek information to compare alternatives before making a decision. Many typologies refer to this concept of rationality in the decision making process, suggesting a coordinated, well-considered approach that examines several possible choices and selects the most favourable option (e.g., Ansoff, 1965; Andrews, 1971).

Figure 3.1: Framework of strategic decision processes

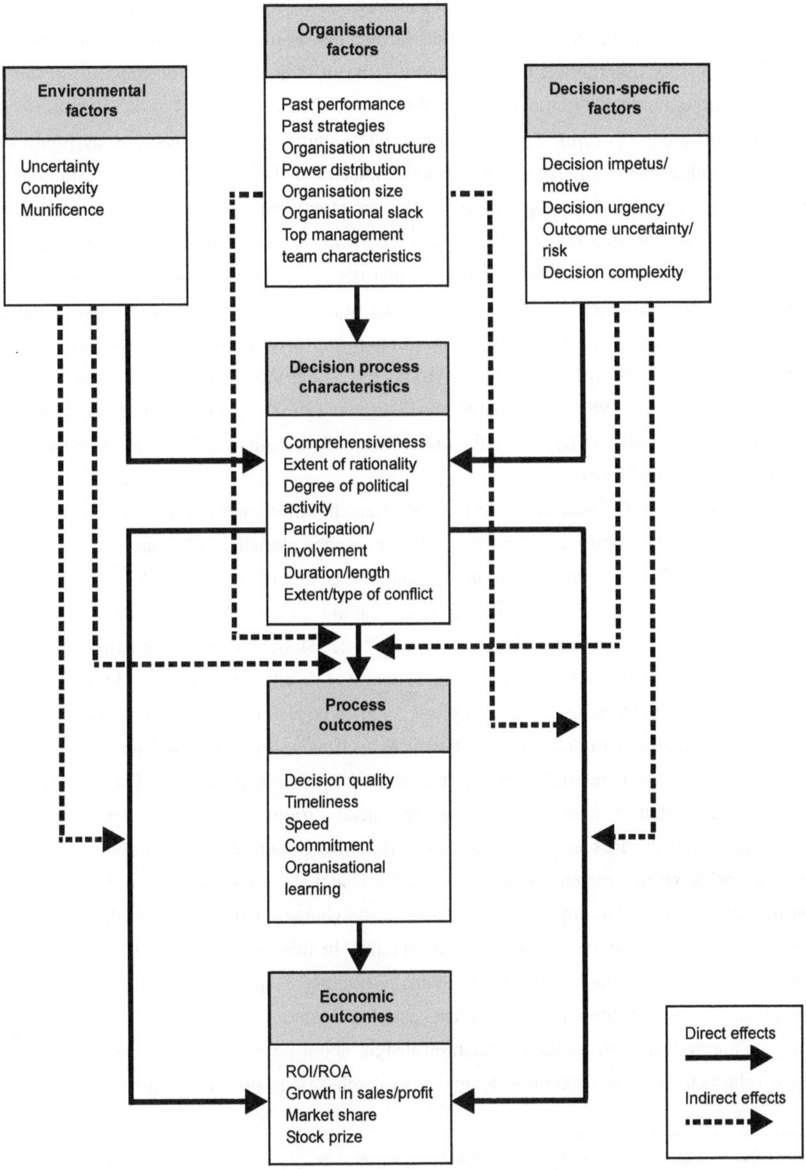

Organisational factors

Past performance
Past strategies
Organisation structure
Power distribution
Organisation size
Organisational slack
Top management
team characteristics

Environmental factors

Uncertainty
Complexity
Munificence

Decision-specific factors

Decision impetus/
motive
Decision urgency
Outcome uncertainty/
risk
Decision complexity

Decision process characteristics

Comprehensiveness
Extent of rationality
Degree of political
activity
Participation/
involvement
Duration/length
Extent/type of conflict

Process outcomes

Decision quality
Timeliness
Speed
Commitment
Organisational
learning

Economic outcomes

ROI/ROA
Growth in sales/profit
Market share
Stock prize

Direct effects

Indirect effects

Source: Rajagopalan, Rasheed, and Datta (1993, p. 352)

In fact, this theoretical model, which does not correspond to many empirical observations (e.g., Bower 1970; Quinn 1980; Burgelman 1983; Pettigrew 1985; Johnson 1987; Hinings and Greenwood 1988; Sminia 1994), has implied further discussions, explanations, and models about the rationality assumption being a central issue in strategic decision making. In their review on strategic decision making, Eisenhardt and Zbaracki (1992) look at empirical support for different concepts, such as bounded rationality and organisations as political systems. Bounded rationality considers decision makers to be rational in some situations while following more heuristic approaches in other situations, especially in environments with a high degree of uncertainty (Eisenhardt and Zbaracki, 1992), dynamism, or complexity (Fredrickson, 1984). The political model states that decisions made within a group might be influenced by contradictory goals and conflicting objectives of the involved individuals. Therefore, the negotiation process and the distribution of power between the individuals have a significant impact on the outcomes (March, 1962; Pettigrew, 1973; Tushman, 1977; Narayanan and Fahey, 1982). Eisenhardt and Zbaracki (1992) concluded that the decision making process is only partly rational and consists of an interaction between bounded-rationality and political processes.

Schwenk (1995) noted four major streams of research within the field of strategic decision making, including (1) strategic decision models and characteristics, (2) biases in strategic decision making, (3) individual and organisational minds, and (4) upper echelons. In the first stream, scholars model the strategic decision making process and categorise strategic decisions, which is complicated because strategic decisions are often unstructured and unorganised. Mintzberg, Raisinghani, and Theoret (1976) differentiated between the identification, development, and selection phase.[8] This area also includes studies dealing with aspects such as rationalism and incrementalism as well as politics and conflicts in strategic decision processes and is related to the discussion in the previous paragraph. The second line of research deals with decisional biases, such as causal attributions, strategic persistence, or biases in recollection.[9] Research that belongs to the third stream attempts to determine how executives perceive the current competitive situation and the environment from a strategic point of view.[10] The last group of studies looks at the characteristics of those players in the organisation who are responsible for strategy making. In this context, scholars have mainly examined top executives and management teams. Generally, research in the field of strategic decision making is dominated by the debate about influence factors on performance (e.g., strategic planning), though it lacks substantial insight about possible antecedents. This study analyses characteristics of executive teams in research-based spin-off companies and their

---

[8] More: Hickson, Butler et al. (1986), Hitt and Tyler (1991), Hart (1992)
[9] Overview: Schwenk (1984, 1986)
[10] More: Huff (1990), Porac and Thomas (1990)

impact on strategic outcomes. Therefore, the next section introduces the upper echelons framework, providing the theoretical background for this study and raising the question regarding the link between management teams and firm level strategy.

3.4   Upper echelons theory

The upper echelons theory (Hambrick and Mason, 1984) states that the characteristics of top managers have an essential influence on the strategic decisions of a company and, furthermore, affect the organisational outcomes (Finkelstein and Hambrick, 1996). This perspective claims that managers have an outstanding impact. The upper echelons theory goes back to cognitive and behavioural research and attempts to explain organisational behaviour by observing the behaviour of the management team. According to the behavioural theory of the firm (Cyert and March, 1963), rationality is not always the basis for management decisions. On the contrary, managers, as human beings, make decisions that are partly influenced by factors such as conflicting objectives and bounded rationality (Nielsen, 2010). The main assumption posits that executives' psychological characteristics, such as values, attitudes, and perceptions, influence the strategic decisions they make. Furthermore, they affect the strategic choices of the organisation and, as a result, the performance. These cognitive constructs, however, are either unobservable or difficult to validate and measure in a reliable way (Pfeffer, 1983). Therefore, the observable demographic characteristics of the management team are taken as proxies reflecting the underlying psychological constructs.

Figure 3.2 shows the original model introduced by Hambrick and Mason (1984). The first box, "The objective situation", on the left-hand side of the model includes internal aspects such as organisational attributes and external factors such as the industry sector, both of which might influence the management team or their strategic choices. The next rectangle, labelled "Upper echelon characteristics", consists of the demographic attributes of the management team, such as age, education, and professional experience, which are observable and can be used as indicators to reflect their cognitive base, including values, perceptions, and attitudes, all of which determine a manager's bounded rationality. The upper echelons theory states that differences in demographic characteristics of firm leaders indicate differences in cognitive processes and their cognitive models, which, in turn, explains variance in firm level strategy. Thus, the third box contains a range of possible variables of strategic outcomes, predominantly including content-related aspects such as diversification, integration, and acquisition. Finally, the model expects executives' characteristics to influence economic outcomes; thus, the last box on the right-hand side shows several performance indicators, such as growth, survival, and profitability.

Figure 3.2: Original upper echelons model

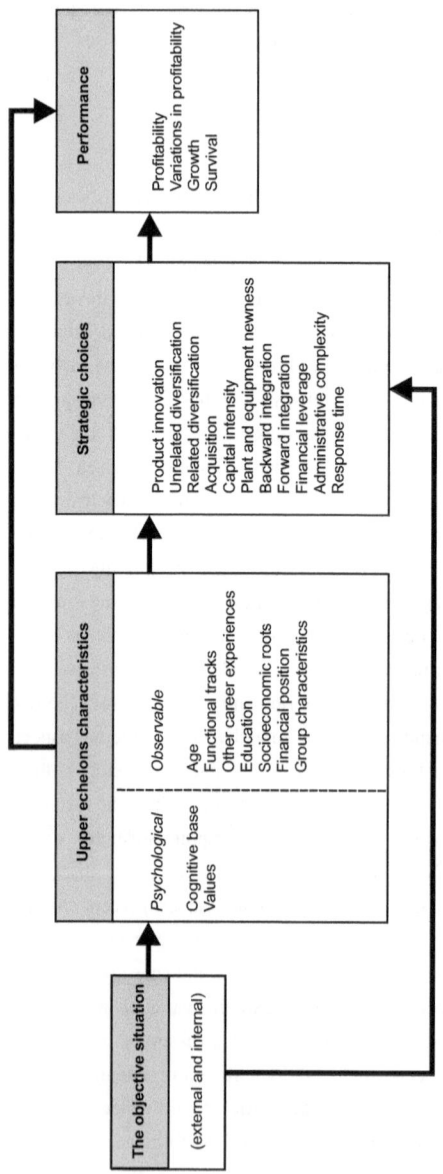

Source: Hambrick and Mason (1984, p. 198)

In summary, strategic choices at the organisational level reflect the cognitive mindset of firm leaders, such as managers or executives. This unobservable cognitive mindset is related to observable attributes such as professional experience. These factors can be used to predict strategy and performance on a firm level.[11] The following example illustrates the theoretical concept. Hambrick and Mason (1984) stated the proposition that young managers are more likely to follow risky strategies than older managers are. The literature provides several arguments for this assumption. For example, older managers are more interested in financial security and, thus, tend to avoid risky strategies (Carlsson and Karlsson, 1970). Hence, the age of the management team can partly predict the strategic orientation of the company.

Similar to the discussion about the definition of an entrepreneurial team, the exact boundaries of the management team are key issues in the strategic management field.[12] Many studies refer to a conceptual definition that is based on Cyert and March (1963) and referred to as dominant coalition. The definitions used in empirical studies differ and may or may not include the board of directors or the advisory board, for instance. These empirical definitions are often based on convenience samples rather than random samples (Carpenter, Geletkancz, and Sanders, 2004). Carpenter, Geletkancz, and Sanders (2004) introduced a second-generation model of the upper echelons perspective that considers previous empirical findings. In contrast to the original model, this framework places more emphasis on contextual factors such as the environmental situation or external situation and organisational aspects such as firm or board characteristics, which may affect the strategy process and the composition of the management team. In line with this argumentation, Keck (1997) postulated, for example, that heterogeneous management teams and executives with a short tenure track are more successful in complex environments, whereas homogeneous teams with a long tenure are better for stable environments. The level of uncertainty is expected to have an impact on the upper echelons model (Keck, 1997), which is presented in figure 3.3. The upper echelons theory is an important and influential framework that helps to understand and explain the relationship between the management team and organisational outcomes. The theory not only focuses on performance measures, but also includes strategic content and process-related aspects. This fact opens up a plurality of possible research questions addressing the effect of the management team on firm performance, thus considering strategy as a type of mediator. A substantial number of studies have referred to this framework and noted that managers are an important factor. Nevertheless, parts of the model are still a black box. For example, there is still only weak empirical evidence regarding the direct effects of demographic attributes on firm performance (Cannella, Finkelstein, and Hambrick, 2008).

---

[11] Reviews of upper echelons literature: Jackson (1992), Finkelstein and Hambrick, (1996), Cannella and Holcomb (2005), Carpenter (2006)
[12] Discussion of the precise definition: Jackson (1992), Pettigrew (1992)

Figure 3.3: Modified model of the upper echelons perspective

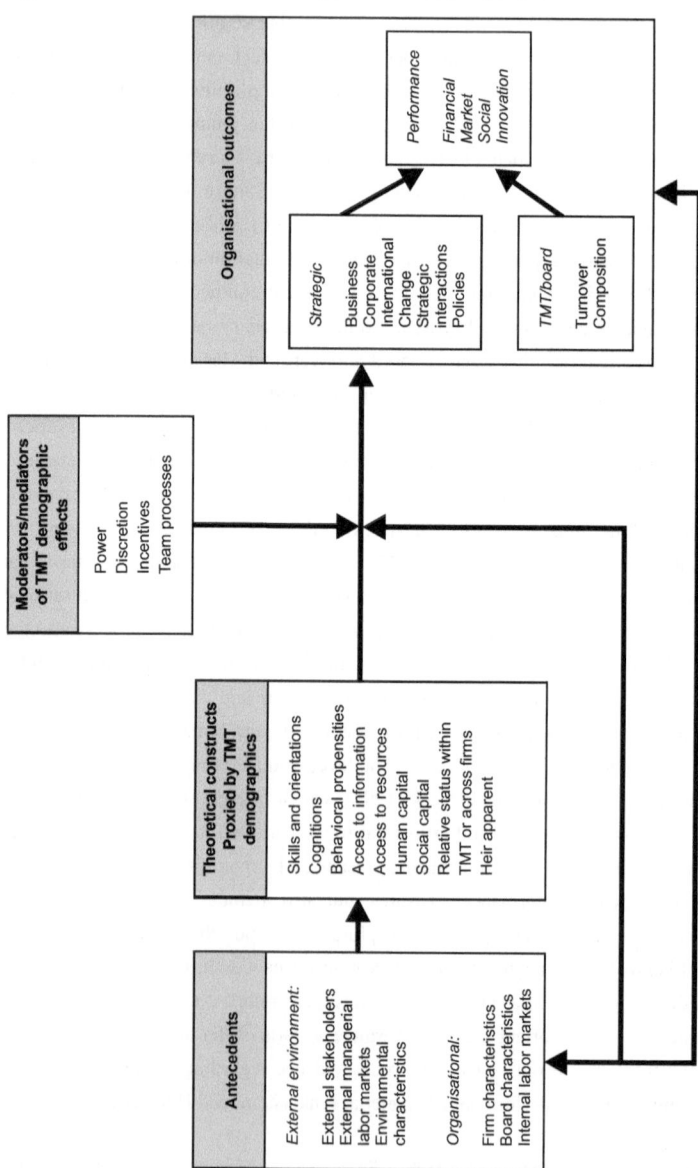

Source: Carpenter, Gelatkancz, and Sanders (2004, p. 760)

3.5   Intended contribution

Although the upper echelons perspective was introduced more in the context of large corporations, applying it in an entrepreneurial setting may add an interesting perspective as it indicates a mediating effect of the characteristics of the management team on firm performance through the team's strategic decisions. This study uses this approach as conceptual framework but applies it to the analysis of the attributes of executive teams in research-based spin-off companies. The upper echelons framework is used for strategic aspects, which are process-related. The unit of analyses of this study are executive teams in academic spin-off companies. As outlined in the previous chapter, the management team is characterised by several specific attributes that differ from other traditional start-ups. The main aspect in this context concerns the composition of the entrepreneurial team, which can consist of researchers from the parent institution or external surrogate entrepreneurs.

This study analyses antecedents and outcomes of strategic decision making using a multi-level approach that considers the team, firm, and environmental levels. The study intends to contribute to the strategic entrepreneurship field by broadening the upper echelons theory in two ways. First, it is applied to high-technology research-based spin-offs, so small and young companies rather than large and established companies are considered. Second, the framework assumes a mediating effect between the characteristics of the management team and the firm performance through their strategic choices. Whereas previous research usually refers to content-related issues, this study considers strategic processes within the company. In summary, in addition to the direct link between management team characteristics and performance, which has been intensively discussed under the umbrella of human capital, this study examines antecedents and consequences of strategic decision making processes. As outlined in the previous sections, the analysis focuses on two main aspects that are crucial in the context of research-based spin-off companies: (1) the level of strategic planning activities and (2) the entrepreneurial strategic posture.

4 Research model

The previous chapter described the theoretical background and introduced the upper echelons perspective as the central framework for this study. This chapter develops the hypotheses and research questions for the empirical analysis. The hypotheses were derived from theoretical and empirical findings, whereas the research questions were formulated if prior studies suggested a relationship between the characteristics of the management team, with a focus on prior experience, strategic decision making, firm performance, organisational attributes, and the environmental situation. Each of the following sections highlights one specific relationship and refers to complementary theoretical approaches, such as the resource-based view of the firm, human capital theory, life-cycle theory, stage-based models, effectuation and causation theory, and the strategic management literature on strategy making. In addition, relevant empirical findings in the context of academic spin-offs, new technology-based ventures, and small and young firms are discussed.

4.1 Examined relationships

The underlying research model for the empirical analysis was derived from the central research question how the characteristics of executives in research-based spin-off companies shape strategy and influence performance. As discussed in the previous chapter, Carpenter, Geletkancz, and Sanders (2004) suggested extending the upper echelons perspective while taking organisational and environmental factors into consideration. Therefore, the following relationships are examined: (1) the link between executive characteristics and firm performance, (2) the link between strategic decision making and firm performance, (3) the link between executive characteristics and strategic decision making, and (4) the link between organisational and environmental aspects and strategic decision making.

The upper echelons perspective is used as the central framework to explain the relationships. In a comprehensive literature review of top management teams, Nielsen (2010) identified the upper echelons perspective as the main theoretical background on which the majority of the examined studies are based. Some scholars have combined the upper echelons theory with other approaches, such as social psychological theories, the resource-based view, agency theory, and entrepreneurship. Previous research about management teams in the context of upper echelons mostly embraced different levels of analysis, such as on the individual, team, firm, or environmental level. Performance measures as outcome variables are a central aspect of this discussion. A combination of the team and firm levels of analysis is common. In most cases, upper echelons studies focus on the management team as the main unit of analysis. As outlined in the previous paragraph, the scope of this study will be extended and will use a multi-level approach that considers the team, firm, and environmental levels. Nielsen (2010) pointed out that a combination of the upper echelons perspective with other theoretical approaches is helpful in explaining and arguing about the directions of specific relationships. Hence, the assumed impact of entrepreneurial teams on strategy and performance in academic spin-off companies will be explained by adding different theoretical lenses. First, this work draws on the resource-based view and human capital theory to explain how initial resources may influence organisational outcomes (Becker, 1964; Wernerfelt, 1984; Barney, 1991). Second, different types and modes of strategy making are discussed including the effects on performance (Miles and Snow, 1978; Venkatraman, 1989). Third, effectuation and causation theory is used to explain why ventures with different executive teams may emphasise different strategic decision processes (Sarasvathy, 2001; Read and Sarasvathy, 2005; Wiltbank, Dew et al., 2006). Finally, this study refers to life-cycle theory to outline the development of research-based spin-offs (Miller and Friesen, 1984; Vohora, Wright, and Lockett, 2004).

Figure 4.1 gives an overview of the considered issues of this study. The first box on the left contains organisational and environmental aspects. The second box shows the examined team characteristics, such as experience, diversity, and composition. In the next box, the discussed strategic decision making processes are presented. Finally, the performance measures are displayed on the right. In the remainder of this chapter, each link will be discussed separately. Every section starts with a brief description of the theoretical lens that is added to the upper echelons framework to explain the relationships. In addition, the empirical findings relating to the specific context are presented in as much depth as possible. Depending on the theoretical and empirical support, either hypotheses or research questions are formulated. The hypotheses postulate a certain direction, whereas the research questions explore the relationships without prior assumptions.

Figure 4.1: Research model

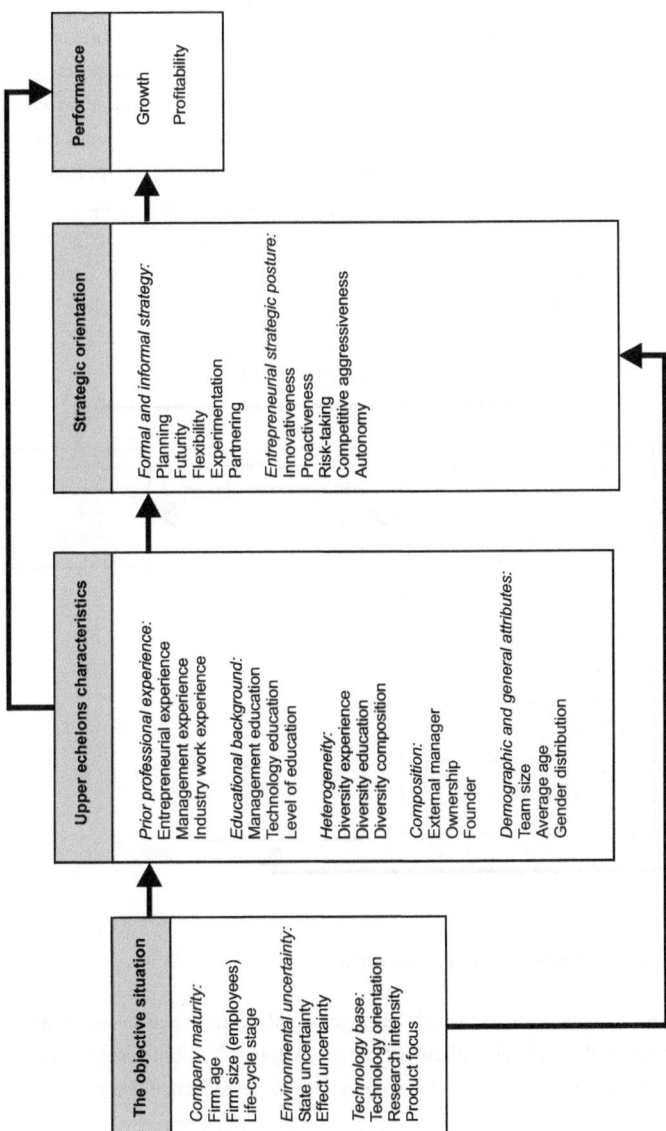

4.2 Link between executive characteristics and firm performance

The first link analysed in this study is the direct impact of management team characteristics on firm performance in research-based spin-off companies. This section discusses prior professional experience, such as entrepreneurial, management, and industry work experience, and educational background, including the main focus and the level of formal qualification, within the executive team. Furthermore, the composition in terms of the ownership, founders, and external managers is considered. An important aspect is the level of heterogeneity, which is studied from different perspectives. Finally, the analysis includes demographical characteristics and general attributes, such as average age, gender, and team size. Figure 4.2 highlights the aspects examined in the following paragraphs.

Figure 4.2: Link between executive characteristics and firm performance

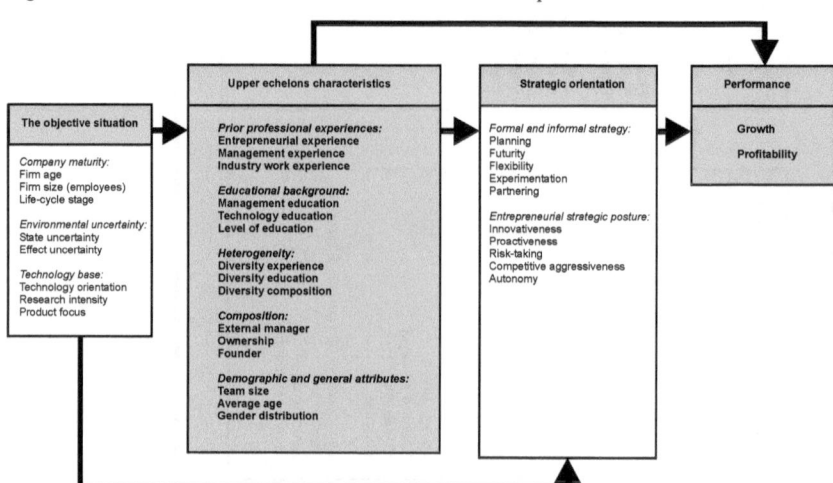

4.2.1 Resource-based view and human capital theory

According to the resource-based theory, which has significantly contributed to the strategic management field and is also critically discussed among scholars[13], initial resource stocks have a significant influence on organisational outcomes such as performance (Barney, 1991).

---

[13] Critical discussion about the resource-based view as a theory: Priem and Butler (2001a, 2001b), Powell (2001), responses: Barney (2001a, 2001b), literature review on resource-based view: Lockett, Thompson, and Morgenstern (2009)

In this context, the ability of a venture to keep and strengthen its competitive position in a market depends on important resources rather than on a specific product market combination. This point of view is especially relevant for research-based spin-off companies because they usually start with a core technology that must be commercially exploited rather than with a well-defined market niche, which should be addressed explicitly. The resource-based view as a framework understands firms as a set of resources that influence performance and competitive advantage. The theory refers to several different theories of the firm and goes back to Edith Penrose and her work "The Theory of the Growth of the Firm" (Penrose, 1959). Literature in the field of strategic management, e.g., Barnard (1938), Selznick (1957), Chandler (1962, 1977), and Rumelt (1974), has discussed the contribution of internal competencies to firm performance and have, thus, already incorporated a resource-based perspective (Conner, 1991). In this context, Wernerfelt (1984) outlined the theory of the resource-based view of the firm with a coherent description and noted, "For the firm, resources and products are two sides of the same coin." (Wernerfelt, 1984, p. 171).

A resource-based approach emphasises the attributes of the venture that are costly to copy and are connected to the organisation. Barney (1986) suggested that this set of resources is significantly responsible for economic rents and, therefore, affects performance and competitive advantage. There are different types of rents concerning the resource-based approach: (1) superior resources that are limited in supply and only available for some companies implicate lower average costs and, thus, so-called Ricardian rents (Peteraf, 1993), (2) barriers to potential competitors imply restrictions of output and result in monopoly rents (Mahoney and Pandian, 1992; Peteraf, 1993), (3) "the difference between the first best and second best use value of a resource" (Mahoney and Pandian, 1992, p. 364) refers to firm-specific resources and leads to quasi rents, and (4) innovations in an uncertain environment may achieve products distinguishable to competitors and entrepreneurial or Schumpeterian rents (Mahoney and Pandian, 1992).

The resource-based view assumes that companies try to obtain above-normal returns. Conner (1991) stated that a company needs either a product or a service different from those of its competitors or the ability to offer the same product at a lower price. According to Rumelt (1984), Barney (1986), and Peteraf (1993), a durable competitive advantage with above-normal returns needs resources that are valuable, rare, imperfectly imitable, and not substitutable. In this context, Peteraf (1993, p. 185) argued the following: "(1) Resource heterogeneity creates Ricardian or quasi-rents. (2) Ex post limits to competition prevent the rents from being competed away. (3) Imperfect factor mobility ensures that valuable factors remain with the firm and that the rents are shared. Resources are perfectly immobile if they cannot be traded. (4) Ex ante limits to competition keep costs from offsetting the rents."

Scholars have introduced several concepts of resource categories that are similar but that differ in key details. Previous research has discussed tangible assets, such as financial capital or physical properties, and intangible resources, such as knowledge, routines, or collaboration activities. Barney (1991) and Brush, Greene, and Hart (2001) established the following often used set of categories: (1) technological resources, (2) social resources, (3) financial resources, and (4) human resources. The term technological resources includes products and technology that belong to the company, such as patented products and services or patented technology (Borch, Huse, and Senneseth, 1999). The portfolio of spin-off companies may rely on a technology base that differs in terms of innovativeness and the development scope or stage. Social resources on the firm level may take into account the firm's relationships and network, for example, to customers and suppliers or to the financial sector (Brush, Greene, and Hart, 2001). Scholars have also used the term social capital in this discussion (e.g., Lee, Lee et al., 2001; Elfring and Hulsink, 2003). The category financial resources reflects the firm's capability to make investments, especially in the early development stages. It covers the form and extent of financial capital, which can either be external, such as venture capital funding, or internal, such as loans, subsidies, and prior profits (Mustar, Renault et al., 2006). Human resources refer to the attributes of the founding team, the personnel, and the current management team. Previous research has used several measures of this term, such as the size and background of the founding team or the composition, diversity, and professional experience of the management team.

The literature referring to the resource-based view indicates that idiosyncratic internal resources implicate a sustainable competitive advantage. The question is which resources are mainly responsible for these relationships in the context of research-based companies. In entrepreneurship research, several studies have identified the founders and their characteristics as having an essential impact on performance and competitive advantage (e.g., Brüderl, Preisendörfer, and Ziegler, 1992; Chandler and Jansen, 1992; Brüderl and Preisendörfer, 2000). Although the discussions of the contributions of individual attributes comprise different arguments and opinions, a number of prior studies (e.g., Wernerfelt, 1984; Dierickx and Cool, 1989; Barney, 1991; Peteraf, 1993; Collis, 1994; Teece, Pisano, and Shuen, 1997) have emphasised the role of the management team in identifying and exploiting profitable opportunities. As early as 1959, Penrose perceived the weight of "unique managerial talent that is inimitable" (Penrose, 1959, p. 35) and distinguished between entrepreneurial and managerial competences.

This study explores the effects of resource endowments on the performance of research-based spin-off companies and focuses on the impact of human capital related aspects. With respect to human resources or human capital, previous research mostly agrees on the relevance of entrepreneurs, owners, and managers as the key resource assets and influence factors for the growth and performance of small companies and, in particular, new ventures (Shane, 2000). The emergence of human capital theory was driven by the intention to analyse how individual investments of employees in training, education, knowledge, and experience improve their personal incomes (Becker, 1964). Scholars from the entrepreneurship field have used human capital theory to explain varying performance levels of newly emerging ventures, for example, compared to more mature companies that are already established in the market. Meanwhile, a substantial number and variety of different measures have been established, such as formal qualification, work experience, and other types of knowledge and capabilities. In the context of young enterprises, task-related variables, such as entrepreneurial, management, or industry work experience, have been discussed as key assets for new venture performance and development (e.g., Sexton and Bowman, 1985; Florin, Lubatkin, and Schulze, 2003). According to human capital theory, individuals can develop their capabilities, leading to higher efficiency and productiveness, if they acquire a wide range of human capital resources (Becker, 1975).

4.2.2   Hypotheses and research questions

In principal, the human capital literature differentiates between general and specific human capital (Becker, 1975). Whereas general human capital can be useful in a broader sense, specific human capital refers to competences and skills that are connected to a specific context. Previous entrepreneurial, management, and industry work experience can certainly add to specific human capital in the entrepreneurship context (Stuart and Abetti, 1990; Cooper, Gimeno-Gascon et al., 1994; Cressy, 1996; Gimeno, Folta et al., 1997; Hsu, 2007) and influence executives' behaviour and performance (Ucbasaran, Alsos et al., 2008). Managers can link their strategic and operational decisions to prior experiences as experts in a certain field. The following sections include aspects that are related to both specific and general human capital.

Prior professional experience:

There is substantial empirical evidence that prior professional experience is positively related to firm performance (e.g., Brüderl, Preisendörfer, and Ziegler, 1992; Gartner, Starr, and Bath, 1999; Aspelund, Berg-Utby, and Skjevdal., 2005). In particular, in the context of academic spin-off companies, prior experience may be a valuable asset (Shane and Stuart, 2002; De Coster and Butler, 2005; Mosey and Wright, 2007) because academics often lack business, industry-specific or entrepreneurial knowledge. Colombo and Grilli (2005) analysed the relationship between the human capital of founders and growth based on a dataset of 506 new technology-based ventures. They concluded that prior work experience is a key influence on growth. Beckman, Burton, and O´Reilly (2007) investigated 161 high-tech firms and observed that diverse prior company affiliations and functional diversity are associated with positive outcomes.

In line with the previous discussion, the following hypotheses concerning the prior professional experience of executives in research-based spin-offs are formulated.

Hypothesis 1.1.1: Prior entrepreneurial experience within the executive team has a positive impact on firm performance.

Hypothesis 1.1.2: Prior management experience within the executive team has a positive impact on firm performance.

Hypothesis 1.1.3: Prior industry work experience within the executive team has a positive impact on firm performance.

Educational background:

Depending on the context, the educational background of the entrepreneurial team can be assessed as either general or specific human capital. Based on a sample of 4,429 small firms, Bates (1990) concluded that owner education partly explains firm longevity. According to Colombo and Grilli (2005), university education in economic and managerial fields and, to a lesser extent, in scientific and technical fields has a positive impact on performance in terms of firm growth. Robinson and Sexton (1994) observed that general education has a strong positive effect on self-employment success. On the contrary, Stuart and Abetti (1990) conducted a survey with 52 new technology-based ventures and found that education beyond a bachelor degree is negatively associated with firm performance. Because research-based spin-off companies are usually established by academics working at the parent organisation, the majority of the entrepreneurs have at least a degree at the bachelor or master level. In addition, in many scientific areas, it is common for the involved researchers to have graduated at the doctoral level. Therefore, the average level of education within the executive team of the spin-off is expected to be very high. Furthermore, scientists tend to be specialists or experts in their fields rather than generalists with a very broad educational profile. Based on this assumption, the proportion of executives with management education is expected to be quite low, whereas the proportion of executives with a degree in a technical field is anticipated to be high.

According to the previous discussion addressing the specific characteristics in terms of the educational background of managers in research-based spin-offs, the following research questions are formulated.

Research question 1.1.1: How is the proportion of executives with an education in management related to firm performance?
Research question 1.1.2: How is the proportion of executives with an education in a technology-oriented field related to firm performance?
Research question 1.1.3: How is the overall level of education of the executive team related to firm performance?

Heterogeneity:

A key aspect in the discussion about the composition of management teams deals with heterogeneity versus homogeneity and the question of how these aspects influence performance. In this respect, studies have made contradictory recommendations for setting up a successful team (Birley and Stockley, 2000). On one hand, a higher degree of heterogeneity in the management team implies a higher level of human capital because it is based on a broader variety of knowledge, capabilities, and skills, which enables the venture to be creative and recognise opportunities more effectively (Guzzo and Dickson, 1996; Williams and O'Reilly, 1998; Ensley and Hmieleski, 2005). On the other hand, homogeneous teams tend to be more cohesive, which in turn leads to fewer affective conflicts and fosters new venture growth (Birley and Stockley, 2000; Ensley, Pearson, and Amason, 2002). Furthermore, team heterogeneity is positively related to turnovers and team departures (Ucbasaran, Lockett et al., 2003; Chandler, Honig, and Wiklund, 2005). Previous research has shown that the composition and the characteristics of the management team have a significant impact on strategic and economic outcomes. In this context, scholars have pointed out that the diversity of certain attributes plays a more important role than their accumulation or general tendency (Blau, 1977; Pfeffer, 1983). The diversity constructs capture the distribution of different characteristics within a group and can include general demographic characteristics and other human capital assets, such as experience or education. However, previous research on management team heterogeneity has supported both a positive impact on company performance (e.g., Weinzimmer, 1997; Aspelund, Berg-Utby, and Skjevdal, 2005; Amason, Shrader, and Tompson, 2006; Beckmann, Burton, and O'Reilly, 2007) and a negative influence (e.g., Ensley, Carland, and Carland, 1998). Furthermore, some studies report no significant relationship between diversity and performance.

Another important facet in this discussion is context. Several scholars claim that heterogeneous teams are more successful in uncertain and dynamic situations because they can base their decisions on a broader range of previous experiences and knowledge. Homogeneous teams are expected to perform better in stable environments (Eisenhardt and Schoonhoven, 1990; Hambrick, Cho et al., 1996; Iaquinto and Frederickson, 1997; Keck, 1997). Priem (1989) examined different types of management team heterogeneity, such as age, education, or functional background, and focused on a stable industry sector. Although he postulated a negative relationship between diversity and firm performance, some findings reveal that, even in stable environments, team heterogeneity can have a positive impact on performance. In summary, the discussed findings are contradictory regarding whether or in which context team diversity is beneficial for companies and how it is related to organisational outcomes. As discussed in the previous chapters, entrepreneurial teams of research-based spin-off companies tend to be more homogeneous than teams of traditional start-ups (Ensley and Hmieleski, 2005) because the team members often have a technology-oriented background. In addition, their performance is significantly lower (Ensley and Hmieleski, 2005). Therefore, including additional competencies such as commercial knowledge or industry-specific experiences indicates team development and might possibly outweigh the negative effects of a higher level of heterogeneity in the executive team, especially in the context of academic spin-offs.

For this reason, with respect to the level of heterogeneity of the executive team in research-based spin-off companies, the following hypotheses are stated.

Hypothesis 1.2.1: The degree of heterogeneity of the executive team in terms of prior professional experience has a positive impact on firm performance.
Hypothesis 1.2.2: The degree of heterogeneity of the executive team in terms of educational background has a positive impact on firm performance.
Hypothesis 1.2.3: The degree of heterogeneity of the executive team in terms of general attributes has a positive impact on firm performance.

Composition:

The next section discusses several aspects regarding the composition of the executive team. Franklin, Wright, and Lockett (2001) examined advantages and disadvantages of surrogate entrepreneurs in the executive team who did not belong to the academic staff and joined the spin-off from outside the parent organisation. They argued that these external managers can be an important source of commercial knowledge and skills, which increase the company's chances of success. The delineation of the boundaries of an entrepreneurial team is difficult. Some scholars have addressed questions regarding ownership and the founding team. Roberts (1991a) questioned the success of academic founders who start working for the spin-off on a part-time basis and keep their university position. Olofsson and Wahlbin (1984) supported this point of view and concluded that spin-offs report higher growth rates if the academics leave the university. Based on a study of 768 young semiconductor firms, Boeker and Karichalil (2002) analysed the consequences of founders leaving their organisations. They observed that founder departure decreases with founder ownership and has a U-shaped relationship with firm growth.

In line with the previous discussion, the following hypothesis and research questions reflect the expected relationships between the composition of the executive team and firm performance in the context of research-based spin-off companies.

Hypothesis 1.3: External managers in the executive team have a positive impact on firm performance.

Research question 1.2.1: How is the proportion of founders in the management team related to firm performance?

Research question 1.2.2: How is the proportion of owners in the management team related to firm performance?

Demographic characteristics and general attributes:

Demographic attributes and general team characteristics might influence firm performance. Most new technology-based firms are established by a team rather than by individuals (Roberts, 1991a). Teams have several advantages because they can rely on a broader human capital base, for example, in terms of knowledge or experience. Song, Podoynitsyna et al. (2008) conducted a meta-analysis based on 31 studies and reported a positive relationship between founding team size and performance. On the contrary, Aspelund, Berg-Utby, and Skjevdal (2005) stated that smaller teams increase the survival probability. Several studies have used the age of managers as a proxy for human capital (e.g., Hellerstedt, 2009) because a higher age is related to greater experience. The average tenure of the executive might reflect the level of experience within the company. As discussed in chapter 2, executive teams of research-based spin-off companies tend to have less female managers than traditional start-ups. Although the upper echelons framework theoretically assumes a link between demographic characteristics and firm performance, empirical evidence in this context is rare (Carpenter, Geletkancz, and Sanders, 2004).

Therefore, regarding the demographic attributes and general team characteristics, the following research questions are formulated.

Research question 1.3.1: How is the size of the executive team related to firm performance?

Research question 1.3.2: How is the average age of the executive team related to firm performance?

Research question 1.3.3: How is the gender distribution of the executive team related to firm performance?

Table 4.1 presents the hypotheses and research questions regarding the link between executive characteristics and firm performance.

Table 4.1: Link between executive characteristics and firm performance

Prior professional experience
Hypothesis 1.1.1: Prior entrepreneurial experience within the executive team has a positive impact on firm performance.
Hypothesis 1.1.2: Prior management experience within the executive team has a positive impact on firm performance.
Hypothesis 1.1.3: Prior industry work experience within the executive team has a positive impact on firm performance.
Educational background
Research question 1.1.1: How is the proportion of executives with an education in management related to firm performance?
Research question 1.1.2: How is the proportion of executives with an education in a technology-oriented field related to firm performance?
Research question 1.1.3: How is the overall level of education of the executive team related to firm performance?
Heterogeneity
Hypothesis 1.2.1: The degree of heterogeneity of the executive team in terms of prior professional experiences has a positive impact on firm performance.
Hypothesis 1.2.2: The degree of heterogeneity of the executive team in terms of educational background has a positive impact on firm performance.
Hypothesis 1.2.3: The degree of heterogeneity of the executive team in terms of general attributes has a positive impact on firm performance.
Composition
Hypothesis 1.3: External managers in the executive team have a positive impact on firm performance.
Research question 1.2.1: How is the proportion of founders in the management team related to firm performance?
Research question 1.2.2: How is the proportion of owners in the management team related to firm performance?
Demographic characteristics and general attributes
Research question 1.3.1: How is the size of the executive team related to firm performance?
Research question 1.3.2: How is the average age of the executive team related to firm performance?
Research question 1.3.3: How is the gender distribution of the executive team related to firm performance?

4.3 Link between strategic decision making and firm performance

The second link in this study deals with different facets of strategic decision making on the organisational level and their impact on firm performance. On one hand, the analysis refers to the aspects of planning and rationality in the decision making process and distinguishes between formal and informal strategy approaches. On the other hand, the effect of a firm having an entrepreneurial strategic posture is discussed, including dimensions of innovativeness, proactiveness, risk-taking, competitive aggressiveness, and autonomy. Figure 4.3 provides a summary of the next section.

Figure 4.3: Link between strategic decision making and firm performance

The objective situation	Upper echelons characteristics	Strategic orientation	Performance
Company maturity: Firm age, Firm size (employees), Life-cycle stage	*Prior professional experiences:* Entrepreneurial experience, Management experience, Industry work experience	*Formal and informal strategy:* Planning, Futurity, Flexibility, Experimentation, Partnering	Growth Profitability
Environmental uncertainty: State uncertainty, Effect uncertainty	*Educational background:* Management education, Technology education, Level of education	*Entrepreneurial strategic posture:* Innovativeness, Proactiveness, Risk-taking, Competitive aggressiveness, Autonomy	
Technology base: Technology orientation, Research intensity, Product focus	*Heterogeneity:* Diversity experience, Diversity education, Diversity composition		
	Composition: External manager, Ownership, Founder		
	Demographic and general attributes: Team size, Average age, Gender distribution		

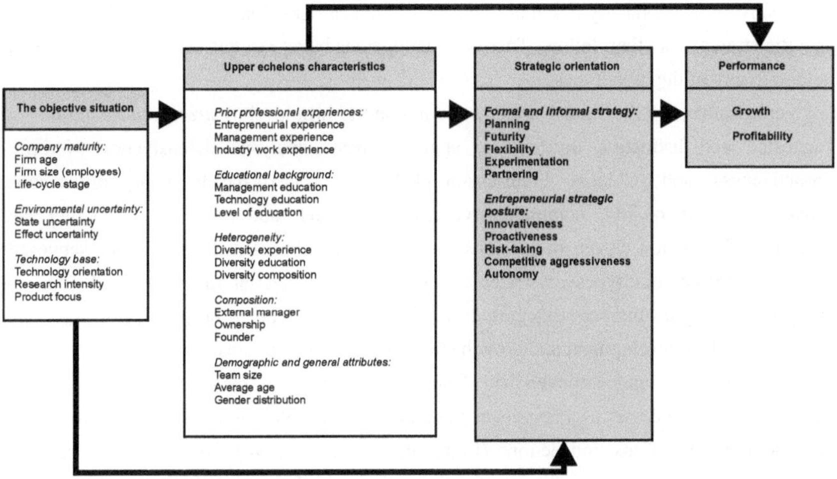

4.3.1 Strategy making modes

Strategy making on the firm level comprises aspects such as decision making (Hart, 1992). In this respect, several frameworks outlining different types and modes of strategy making have been introduced. Hart (1992) distinguished five modes of strategy making: command, symbolic, rational, transactive, and generative. Miles and Snow (1978) differentiated four types, which they termed prospector, defender, analyser, and reactor. They claimed that different strategies on the firm level result from their decisions on how to face three key problems: entrepreneurial, engineering (operational), and administrative. The entrepreneurial problem addresses the market share, the engineering problem regards the implementation of their specific solution, and the administrative problem focuses on structural issues. Prospectors try to explore and exploit new opportunities by developing new products and introducing new technologies to the market. They have a strong focus on innovation and try to change the industry and create new markets. Thus, on an operational level, they need to figure out how to avoid being dependent on only one technology. From an administrative point of view, these organisations have a low level of formalisation. In contrast, companies belonging to the defender type try to stabilise their market share and keep their position. There is little initiative to explore opportunities; they would rather focus on cost efficiency and specialise in a small range of products and services. On the administrative level, they establish formal processes and use long-term planning. Analyser organisations can be seen as a mixture of prospectors and defenders. Although they are interested in exploiting new opportunities, their own innovations are mostly based on imitation to reduce risks. Finally, reactors do not have a specific strategy that they follow. Therefore, companies belong to this category if they do not belong to any of the other types.

Venkatraman (1989) framed strategy making on the firm level as *strategic orientation* and suggested six distinctive dimensions: aggressiveness, analysis, defensiveness, futurity, proactiveness, and riskiness. Fredrickson (1986) mentioned the following dimensions: proactiveness, rationality, comprehensiveness, risk-taking, and assertiveness. Miller and Friesen (1978) found eleven dimensions for strategy making. Borch, Huse, and Senneseth (1999) identified four types of strategic orientations: managerial firms are predominantly analysers and apply market strategies, technological firms tend to be prospector types that focus on product development and growth strategies, traditional firms usually carefully handle risks and growth, and impoverished firms do not have a clear strategic orientation. Entrepreneurial processes as a firm level phenomena refer to decision making practices, styles and methods, intentions and actions (Lumpkin and Dess, 1996). An entrepreneurial firm corresponds predominantly to prospector firms (Miles and Snow, 1978) or entrepreneurial organisations (Mintzberg, 1973), whereas non-entrepreneurial or conservative (Covin and Slevin, 1989) companies are similar to defender firms (Miles and Snow, 1978) or adaptive organisations (Mintzberg, 1973).

4.3.2 Hypotheses and research questions

Despite progress in the field of strategic decision making, several questions still remain, such as how certain processes influence performance. As early as Penrose (1959), a differentiation was made between "managerial competence" and "entrepreneurial competence", and it has been noted that many small firms are managerially rather than entrepreneurial oriented. As outlined in the previous chapter, this study focuses on the following questions: should executives in research-based spin-off companies act entrepreneurially, plan strategically, or both? The first approach includes more planning-related aspects, whereas the latter focuses more on risk-taking and opportunity exploration, which determine an entrepreneurial posture. Therefore, the first part of the following section outlines the assumed impact of strategic planning activities on firm performance, whereas the second part deals with the consequences of an entrepreneurial posture.

4.3.2.1 Formal and informal strategy making

Rationality in the decision making process proposes a coordinated, well-considered approach that compares different options and selects the most promising alternative. Thus, strategic planning is a key aspect of the strategy process research stream. In the strategic management field, many scholars have focused on the relationship between planning and performance (e.g., Armstrong, 1982; Shrader, Taylor, and Dalton, 1984; Pearce, Freeman, and Robinson, 1987), and a considerable number of studies have reported evidence for a positive relationship between strategic planning and firm performance (e.g., Baker, Addams, and Davis, 1993; Miller and Cardinal, 1994; Hopkins and Hopkins, 1997; Berry, 1998; Andersen, 2000, 2004). The link between planning and performance has been extensively discussed within the scope of large corporations, and the studies predominantly indicate a positive influence of strategic planning activities (e.g., Bracker, Keats, and Pearson, 1988; Bracker and Pearson, 1988; Schwenk and Shrader, 1993).[14] However, few empirically based quantitative studies focus on small companies (Kraus, Harms, and Schwarz, 2006).

As mentioned in the previous chapters, research-based spin-offs as a subgroup of new technology-based ventures must cope with liabilities such as newness and uncertainty. Several studies have focused on companies in uncertain and dynamic environmental situations. Comprehensiveness as a measure of rationality in decision processes has been found to have a negative (e.g., Fredrickson and Mitchell, 1984) and a positive influence (e.g., Eisenhardt, 1989) on firm performance with regard to companies in an unstable (Fredrickson and Mitchell, 1984, p. 399) or high-velocity environment (Eisenhardt, 1989). Priem, Rasheed, and Kotulic (1995) reported a positive relationship between rationality and performance for firms operating in dynamic environments but not for companies in a stable environment. Several scholars have found empirical evidence for a positive relationship between formal planning and firm performance in small and medium-sized companies and in new ventures. Schwenk and Shrader (1993) conducted a meta-analysis that looked at 14 studies on strategic planning

---

[14] More discussion including inconsistent findings: Brews and Hunt (1999), Hutzschenreuter (2006)

and financial performance in small firms. The overall findings suggest a significant positive link between planning and performance. Based on a sample of 217 small and medium-sized companies in the electronics industry, Bracker and Pearson (1988) observed that structured strategic planners outperform other companies that do not apply formalised planning procedures. Robinson, Salem et al. (1986) found a positive link between profitability and specific analytical planning methods. Sexton and van Auken (1985) noted that future-oriented resource planning contributes positively to firm performance. Smith (1998) concluded that strategic planning is beneficial, especially for very young firms. Schröder (2008) studied 1,803 new technology-based ventures and reported a positive relationship between an analytical and long-term oriented planning approach and firm performance.

Referring to the previous discussion, the following hypotheses are stated in the context of research-based spin-off companies.

Hypothesis 2.1: The level of formal strategic planning activities has a positive impact on firm performance.

Hypothesis 2.2: The level of long-term firm orientation has a positive impact on firm performance.

In contrast to the approaches focusing on rational decision making, scholars have outlined alternative concepts because, especially in the context of small firms and new ventures in an early stage, the benefits of formal planning procedures are still controversial (Berry, 1998; Brinckmann, Grichnik, and Kapsa, 2010). Slevin and Covin (1997) distinguished between planned and emergent strategy modes and observed that the environment and organisational attributes affect the strategy-performance relationship. Lumpkin and Dess (1995) stated that less complex approaches in the strategy making process may be useful for firms in an early development stage but have a negative impact in later phases when the organisation becomes more complex. In this context, Sarasvathy (2001) differentiated between rational approaches of decision making and processes that are more based on intuition and heuristic-based logic. These two perspectives are framed as *causation* and *effectuation*. Whereas the first approach corresponds to the aspects discussed in the previous sections, the latter concept emphasises informal and incremental decision making processes to a greater extent. This approach encompasses key aspects such as experimentation or flexibility.

Therefore, the following research question in the context of research-based spin-off companies is formulated.

Research question 2: How does informal strategy making influence firm performance?

4.3.2.2 Entrepreneurial strategy making

An entrepreneurial strategic posture reflects the attitude of an organisation towards entrepreneurial decisions and actions (Lumpkin and Dess, 1996; Wiklund and Shepherd, 2003) and is also termed as entrepreneurial orientation (Lumpkin and Dess, 1996; Lyon, Lumpkin, and Dess, 2000; Lumpkin and Dess, 2001; Dess, Gregory, and Lumpkin, 2005). The entrepreneurial orientation construct originated from research on strategy making processes (e.g., Mintzberg, 1973) and refers to a strategic choice perspective (Child, 1972), for example, in the context of new entry decisions. Entrepreneurial behaviour on the organisational level has also been discussed by referring to concepts such as entrepreneurial posture (e.g., Covin and Slevin, 1989, 1990, 1991; Zahra, 1993; Knight, 1997), entrepreneurial management (e.g., Stevenson and Jarillo, 1990; Brown, Davidsson, and Wiklund, 2001), corporate entrepreneurship (e.g., Zahra and Covin, 1993; Hornsby, Kuratko, and Zahra, 2002), and firm entrepreneurship (e.g., Miller and Friesen, 1982; Miller, 1983).

The key dimensions of the entrepreneurial orientation construct go back to Miller (1983, p. 771): "An entrepreneurial firm is one that engages in product market innovation, undertakes somewhat risky ventures, and is first to come up with 'proactive' innovations, beating competitors to the punch". Thus, the three dimensions of *innovativeness*, *risk-taking*, and *proactiveness* describe an entrepreneurial posture as a firm level phenomenon. Lumpkin and Dess (1996) argued for two additional dimensions: *autonomy* and *competitive aggressiveness*. Meanwhile, the entrepreneurial orientation concept has been analysed by several scholars, so it is one of the areas in the entrepreneurship field for which extensive knowledge and a broad body of literature could be developed (Rauch, Wiklund et al., 2009). Prior empirical studies have shown a significant positive relationship between entrepreneurial orientation and firm performance. These findings are robust and consistent across different industry sectors, conceptualisations, and countries. Rauch, Wiklund et al. (2009) conducted a meta-analysis focusing on the relationship between entrepreneurial orientation and firm performance. Based on the 51 studies that were taken into consideration, an average correlation of $r = 0.242$ was reported.

Microbusinesses reported a larger correlation than small businesses, whereas no differences could be found between micro and large firms or between small and large companies (Rauch, Wiklund et al., 2009). Because technological shifts and breakthroughs occur more dramatically and quickly in high-tech industries, the greater impact of an entrepreneurial posture on firm performance seems to be self-explanatory. Furthermore, entrepreneurial orientation has a stronger connection to firm performance in emerging industries. When the industry sector becomes more mature, the relationship weakens and can even invert (Covin and Slevin, 1990). In this context, an entrepreneurial orientation appears to be highly relevant for research-based spin-off companies that operate in high-tech industries and face a high degree of uncertainty regarding the environmental situation.

However, entrepreneurial orientation has predominantly been discussed as a unidimensional construct. Because most of the different dimensions correlate significantly on a high level to one another (Rauch, Wiklund et al., 2009), studies often form one unidimensional factor (e.g., Covin and Slevin, 1989, 1990; Wiklund and Shepherd, 2003, 2005; Walter, Auer, and Ritter, 2006). Nevertheless, some scholars have argued that the dimensions of the entrepreneurial orientation construct describe different facets of the phenomenon (e.g., Lumpkin and Dess, 1996, 2001; Kreiser, Marino, and Weaver, 2002; Naldi, Nordquist et al., 2007). For this reason, entrepreneurial orientation should be understood as a multidimensional construct, and each dimension may have a different impact on firm performance depending on the contextual situation. Thus, according to the previous discussion, the following hypothesis is stated.

Hypothesis 2.3: Innovativeness, proactiveness, risk-taking, competitive aggressiveness, and autonomy are distinct dimensions of the entrepreneurial orientation concept.

In the following section, each dimension of the entrepreneurial orientation construct is discussed separately.

Innovativeness:

The crucial role of innovations in the entrepreneurial context has already been mentioned by Schumpeter (1934). The innovativeness dimension describes companies that are open to new ideas and rely on new technologies that differ from common practices to improve such processes and to develop new products or services (Damanpour, 1991; Pearce, Kramer, and Robbins, 1997; Hurley and Hult, 1998). Thus, such firms emphasise R&D and foster creativity and experimentation to generate new combinations, to gain technological leadership, and for novelty purposes. Zaltmann et al. (1973, p. 64) stated that "openness to innovation" describes whether there is a consensus within the organisation about adopting new ideas. Empirical evidence suggests a positive relationship between innovativeness and firm performance (e.g., Deshpande, Farley, and Webster, 1993; Kreiser, Marino, and Weaver, 2002), especially if firms are effectively adopting the innovations generated within the organisation and bringing them to the market (Damanpour, 1991). Zahra (1996) argued that innovativeness is crucial for a company to survive and stated that "success in today's competitive environment requires a company to pursue a coherent technology strategy to articulate its plans to develop, acquire, and deploy technological resources to achieve superior financial performance" (Zahra, 1996, p. 189). Innovativeness seems to be a major aspect for research-based spin-off companies because they often rely on new knowledge and technologies. Therefore, the following hypothesis is stated:

Hypothesis 2.4.1: The level of innovativeness is positively associated with firm performance.

Proactiveness:

Proactiveness characterises companies that are continuously looking for new market opportunities to exploit and capitalise. They take the initiative to meet emerging customer needs and respond to changing demands, which is especially relevant in growing markets. By being first, proactive ventures try to realise a *first mover advantage* (Dess, Lumpkin, and Covin, 1997; Lieberman and Montgomery, 1998). They may shape the characteristics of the future environment and competition in a beneficial way for the company. In this respect, the dimension differs from innovativeness, which more strongly emphasises innovation and novelty. Nevertheless, proactiveness and innovativeness are very likely to be correlated to each other because initiating behaviour plays a major role for both. Furthermore, the proactiveness construct is related to the following concepts: prospector type (Miles and Snow, 1978), prospector strategy (Davig, 1986), and prospector orientation (Aragón-Sánchez and Sánchez-Marín, 2005). Previous empirical findings have suggested a positive relationship between proactiveness and firm performance, especially in the context of technology-oriented companies (e.g., Davig, 1986; Miller and Toulouse, 1986) and young businesses (e.g., Kotey and Meredith, 1997; Wiklund and Shepherd, 2005). Thus, the following hypothesis is formulated.

Hypothesis 2.4.2: The level of proactiveness is positively associated with firm performance.

Risk-taking:

Already Cantillon (1755) considered taking risks as a key characteristic of entrepreneurship. Therefore, the risk-taking dimension is an essential part of the entrepreneurial orienttation construct and specifies an aspect of strategic behaviour. These companies show a strong tendency and willingness to take high risks to achieve high returns. This dimension is more intuitive than analytical. Thus, bold actions can lead to possible profits and potential losses (Morgan and Strong, 2003). Begley and Boyd (1987) report a curvilinear relationship between risk-taking and firm performance, stating that "risk-taking has a positive effect on ROA up to a point. Beyond that point, increases in risk-taking began to exert a negative effect on ROA"

(Begley and Boyd, 1987, p. 89). However, empirical evidence suggests a positive relationship between risk-taking and performance in the context of new technology-based ventures (Schröder, 2008) and small firms (Miller and Toulouse, 1986). For this reason, the following hypothesis is postulated.

Hypothesis 2.4.3: The level of risk-taking is positively associated with firm performance.

Competitive aggressiveness:

The competitive aggressiveness dimension describes the firm's ambition to achieve a superior position in the market and to beat its rivals directly or respond offensively to their moves. They invest substantial resources to gain a higher market share because doing so is supposed to increase influence and profits in the long term. Davidson (1987, p. 161) outlined the offensive character of this approach in which companies carry out "direct frontal attacks to drive or overwhelm a competitor". The dimension differs from proactiveness because it focuses on growth in an existing market, so the expansion of a firm's market share is at the expense of the competitor's. Prior research reports empirical evidence for a positive relationship between aggressiveness and firm performance, especially in fast-moving and highly competitive product markets (Covin and Slevin, 1991; Zahra, 1993) and for young companies (Biggadike, 1979; McDougall and Robinson, 1990; Kotey and Meredith, 1997). According to the previous discussion, the following hypothesis is formulated.

Hypothesis 2.4.4: The level of competitive aggressiveness is positively associated with firm performance.

Autonomy:

Autonomy reflects how firms emphasise and foster independent action within the organisation. It can include both individuals and teams that are able to act freely to a certain degree to identify opportunities and bring forward business ideas. Few studies in the area of new technology-based ventures or research-based spin-off companies specifically address the autonomy dimension of the entrepreneurial orientation construct as an independent facet and link it to firm performance. George, Wood, and Khan (2001) included the autonomy aspect in a unidimensional scale measuring entrepreneurial orientation and found a positive link to performance in a study of 70 banks. For this study, the following hypothesis is assumed.

Hypothesis 2.4.5: The level of autonomy is positively associated with firm performance.

Table 4.2 provides an overview of the discussed hypotheses and research question regarding the link between strategic decision making and firm performance.

Table 4.2: Link between strategic decision making and firm performance

Formal and informal strategy making
Hypothesis 2.1: The level of formal strategic planning activities has a positive impact on firm performance. *Hypothesis 2.2: The level of long-term firm orientation has a positive impact on firm performance.* *Research question 2: How does informal strategy making influence firm performance?*
Entrepreneurial strategy making
Hypothesis 2.3: Innovativeness, proactiveness, risk-taking, competitive aggressiveness, and autonomy are distinct dimensions of the entrepreneurial orientation concept. *Hypothesis 2.4.1: The level of innovativeness is positively associated with firm performance.* *Hypothesis 2.4.2: The level of proactiveness is positively associated with firm performance.* *Hypothesis 2.4.3: The level of risk-taking is positively associated with firm performance.* *Hypothesis 2.4.4: The level of competitive aggressiveness is positively associated with firm performance.* *Hypothesis 2.4.5: The level of autonomy is positively associated with firm performance.*

4.4 Executive team characteristics and strategic decision making

Whereas the previous paragraph discussed possible outcomes, the following section outlines antecedents of strategic processes. Rajagopalan, Rasheed, and Datta (1993) distinguished different categories of factors that influence the decision making process. On one hand, they identified environmental factors such as uncertainty, munificence, dynamism or complexity. On the other hand, organisational attributes, such as firm structure, size or age, technology, and characteristics of the management team, such as size or heterogeneity, were also mentioned. Research on the link between characteristics of the management team and strategy making is rare. This aspect will be addressed in the following section and is outlined in figure 4.4.

Figure 4.4: Link between executive characteristics and strategic decision making

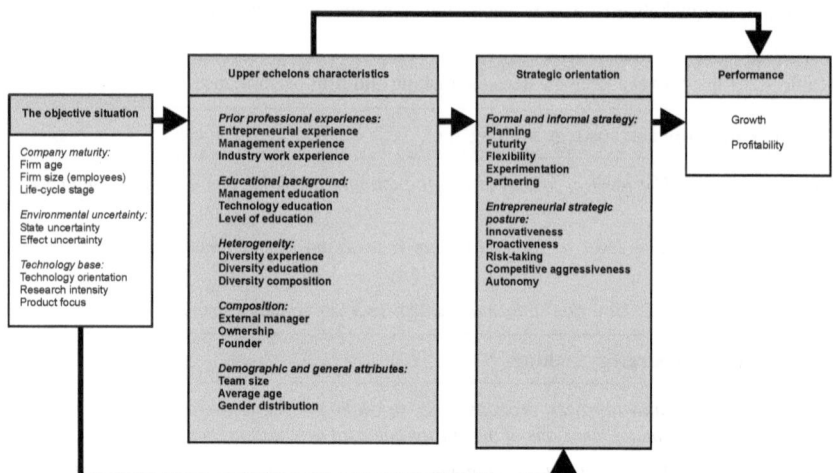

4.4.1   Effectuation and causation theory

Effectuation and causation theory (Sarasvathy, 2001) is rooted in the field of new venture development and addresses how entrepreneurs make strategic decisions in an uncertain and dynamic environment. Research-based spin-off companies that rely on scientific results and new technology often face a similar situation. They have developed innovative products that must be introduced in the market and sometimes even create new markets. Spin-offs usually start with a small entity but can potentially grow very quickly. Furthermore, changes in the management team are very likely due to the absence of commercial knowledge and skills. This fact leads to the key question of how companies should deal with uncertainty to sustain or achieve competitive advantage. The strategic management literature referring to planned and rational or adaptive approaches provides a variety of answers. Scholars representing either the planning or learning school suggest different approaches. The planning division states that companies must put more emphasis on analysing and information gathering when the level of uncertainty increases to outperform their competitors. Empirical studies support this point of view (e.g., Miller and Friesen, 1983; Pearce, Robbins, and Robinson, 1987; Miller and Cardinal, 1994; Priem, Rasheed, and Kotulic, 1995; Dean and Sharfman, 1996; Goll and Rasheed, 1997; Brews and Hunt, 1999). On the contrary, representatives of the learning school argue that ventures must focus more on learning, experimenting, and making incremental steps forward to be able to adapt quickly to changing situations. This perspective is also supported by empirical research (e.g., Fredrickson and Mitchell, 1984; Fredrickson and Iaquinto, 1989; Miller, 1993; Grant, 2003; Hough and White, 2003).

Wiltbank, Dew et al. (2006) broadened the discussion between the planning and learning school, or the planning and adaptive approaches, respectively, by adding a new perspective and disentangling the concepts of prediction and control, which are linked in the traditional strategic management literature. Along this line, assumptions about the environment can be made along two independent dimensions. According to the first dimension, the environment can be seen or perceived as being either predictable or unpredictable. In a predictable environment, putting effort into forecasting seems useful for predicting the future competitive situation and successfully positioning the company. Looking at the environment through an entrepreneurship lens opens another perspective because entrepreneurship is about developing, exploring, identifying, and exploiting new opportunities that can actually affect the environment if new products or technologies successfully reach the market. The second dimension differentiates between an exogenous environment in which business opportunities are given and an endogenous environment in which opportunities can partly be created by the entrepreneur, in other words, the situation in which the environment can be shaped and that in which it cannot be shaped by the venture itself so that entrepreneurs have control over events in their environment. Based on this differentiation between prediction and control in terms of strategy making, Wiltbank, Dew et al. (2006) delineated four categories and related the recent literature in strategic management to them.

The first category refers to a predictable environment that cannot be changed or controlled. Wiltbank, Dew et al. (2006) labelled this fraction as *planning strategies* that focus on forecasting and planning (e.g., Ansoff, 1979; Porter, 1980; McGrath, 1999; Schoemaker, 2002). Second, within a dynamic and unpredictable but exogenous determined environment, so-called *adaptive strategies* emphasise more flexibility and short-term planning (e.g., Quinn, 1980; Eisenhardt, 1989; Mintzberg, 1994b; Teece, Pisano, and Shuen, 1997). These two types are also called *positioning strategies*, whereas the following belong to the *construction strategies*. This term means that future environmental settings can be constructed or influenced; the entrepreneur is seen to be in charge and has control. In the case of a predictable situation, entrepreneurs follow so-called *visionary strategies*. They try to realise their vision (e.g., Hamel and Prahalad, 1991; Courtney, Kirkland, and Viguerie, 1997; Rindova and Fombrun, 1999; Tellis and Golder, 2002). The last category comprises *transformative strategies* that focus on new goals that might shape the future environment, which cannot be predicted (Hayes, 1985; Kim and Mauborgne, 1997; Sarasvathy, 2001).

Sarasvathy (2001) introduced the two alternative concepts of causation and effectuation, which reflect strategic decision making in the context of new venture creation. The approach highlights that entrepreneurs in the venture creation stage might act differently than entrepreneurs or managers in later phases. Whereas causational reasoning in the decision making process corresponds to the concepts of rational and planned strategy (e.g., Brews and Hunt, 1999), the recently introduced concept of effectuation is similar to emergent strategies (e.g., Mintzberg and Waters, 1985) and assumes that the entrepreneurs emphasise informal aspects, such as flexibility and experimentation, more extensively, especially in an uncertain environmental situation.

Causation is based on the approach of prediction and effectuation on the approach of control. In this context, Sarasvathy (2001, p. 245) provides the following definition: "Causation processes take a particular effect as given and focus on selecting between means to create that effect. Effectuation processes take a set of means as given and focus on selecting between possible effects that can be created with that set of means." If the level of analysis is the firm, these means are – according to the resource-based view – equivalent to the physical, human, and organisational resources of the company. Causation corresponds to strategy approaches that are theoretically based on rational decision making and focus on activities such as planning and analysing to achieve a competitive advantage (e.g., Mintzberg, 1978; Ansoff, 1979; Brews and Hunt, 1999). In contrast, effectuation processes are related to emergent and non-predictive strategies and rely more on aspects such as flexibility and experimentation. The theoretical background refers to cognitive science.[15]

4.4.2 Hypotheses and research questions

The main interest of this study focuses on the specific characteristics of executive teams in research-based spin-off companies. This area embraces the composition of the team, which consists of academic entrepreneurs and, in all likelihood, surrogate entrepreneurs from outside the parent institution. The potential lack of commercial knowledge and prior professional experience in particular in a business environment is a central aspect that should be addressed. Thus, the hypothesis development focuses on these issues and refers to effectuation and causation theory to explain how prior experiences, expert knowledge, and team composition are expected to influence strategic decision making processes. Research questions address the remaining aspects, such as educational background or demographic attributes.

---

[15] Review of research on entrepreneurial effectuation by Perry, Chandler, and Markova (2011).

Prior professional experience:

The basic assumption is that executives' experience influence their cognition and understanding (Finkelstein, Hambrick, and Cannella, 2008). In particular, in complex, difficult, and unclear situations, decision makers rely on prior experiences, referring to behaviour patterns that were positively approved in the past (Cyert and March, 1963). Empirical evidence regarding the effect of previous experiences on the intensity of strategic planning is limited. Brouthers, Brouthers, and Werner (2000) concluded that managers of small companies are more likely to be rational in gathering information but are more intuitive in the strategic decision making process. Wally and Baum (1994) mentioned that the speed of how strategic decisions are made is affected by the characteristics of the management team. Along this line, Forbes (2005) stated that experienced entrepreneurs tend to make quicker decisions. Previous entrepreneurial, management, and industry work experience influences executives' behaviour and performance because they can link their strategic and operational decisions to prior experiences, as experts in a field do. Therefore, they are more likely to rely on intuitive reasoning, apply heuristic-based logic, and gather information in a more flexible and selective approach, especially in a dynamic and uncertain environment (Spence and Brucks, 1997; Mitchell, Busenitz et al., 2007). Entrepreneurs and managers without any prior experience do not have these reference points; thus, they act and search for information more systematically and analytically (Gustaffson, 2006). Dew, Read et al. (2009) analysed the decision making process by comparing a group of expert entrepreneurs and a group of students. They concluded that experienced entrepreneurs reason more analogically and think more holistically. Thus, they are generally more likely to use effectual logic than entrepreneurs without any previous experience. As outlined above, effectual reasoning is consistent with non-planned strategy approaches.

Thus, regarding the prior professional experience of executive teams in research-based spin-off companies, the following hypotheses and research question are stated.

Hypothesis 3.1.1: Previous professional experience of the executive team affects strategic planning activities.

Hypothesis 3.1.2: Executive teams with prior entrepreneurial, management, or industry work experience emphasise less formal strategic planning activities.

Hypothesis 3.1.3: Executive teams with prior entrepreneurial, management, or industry work experience emphasise more informal strategic planning activities.

Research question 3.1: How does prior entrepreneurial, management, or industry work experience within the executive team influence the entrepreneurial strategic posture?

Educational background:

Prior studies have suggested that the educational background of the executive team influences their individual values and perceptions and is reflected in certain attributes on the organisational level. Empirical evidence supports a positive relationship between an executive's education and innovativeness (e.g., Thomas, Litschert, and Ramaswamy, 1991). For example, Tyler and Steensma (1998) mentioned that managers with an engineering degree are more receptive to technology opportunities than other executives. Barker and Mueller (2002) examined larger companies and observed that firms with a higher proportion of executives with an MBA degree focus less on R&D. Bertrand and Schoar (2003) also analysed executive teams in large corporations and concluded that "CEOs with MBA education appear to follow more closely the textbook guidelines when making investment decisions" (2003, p. 1203). In this context, Hambrick and Mason (1984, p. 201) stated that managers with an MBA degree are more likely to be "organizers and rationalizers".

In summary, there is empirical evidence that the educational background of management teams influences their strategic decision behaviour. For this reason, the following research questions are formulated.

Research question 3.2.1: How does the educational background of the executive team influence formal strategic planning activities?

Research question 3.2.2: How does the educational background of the executive team influence informal strategic planning activities?

Research question 3.2.3: How does the educational background of the executive team influence the entrepreneurial strategic posture?

Heterogeneity:

In previous studies, heterogeneity within the executive team has often been associated with a higher level of creativity and more effectiveness in decision making (Bantel and Jackson, 1989). On the other hand, some scholars have stated that a high level of heterogeneity may lead to less communication in the management team and lower performance in decision making (O'Reilly, Snyder, and Boothe, 1993). Empirical studies reported inconsistent findings (e.g., Bantel and Jackson, 1989; Murray, 1989; Jackson, Brett et al., 1991; Michel and Hambrick, 1992; Wiersema and Bantel, 1992; Glick, Miller, and Huber, 1993; Smith, Smith et al., 1994). Miller, Burke, and Glick (1998) studied the impact of heterogeneity in the executive team on the comprehensiveness of strategic decision making and the extensiveness of strategic planning. They concluded that diversity has a negative effect on extensive long-term planning. Research-based spin-off companies often start with a homogeneous management team, especially if they consist exclusively of former staff members of the academic institution. However, new members such as surrogate entrepreneurs are likely to join the team to fill existing gaps. In line with Miller, Burke, and Glick (1998) and the argument in the previous section regarding the impact of prior experiences on strategic planning activities, the following hypotheses and research question are stated.

Hypothesis 3.2.1: Heterogeneity in the executive team affects the intensity of strategic planning activities.

Hypothesis 3.2.2: Heterogeneous teams emphasise less formal strategic planning activities than homogeneous teams.

Hypothesis 3.2.3: Heterogeneous teams emphasise more informal strategic planning activities than homogeneous teams.

Research question 3.3: How does heterogeneity within the executive team influence the entrepreneurial strategic posture?

Composition:

The composition of the management team may influence strategic decision making. As outlined in the previous chapters, research-based spin-off companies may include external managers and surrogate entrepreneurs from outside the research institute to bridge managerial constraints. The contribution of professional executives with a managerial mindset is supposed to change the strategic decision making. Based on a sample of 295 small and medium-sized companies, Escribá-Esteve, Sánchez-Peinado, and Sánchez-Peinado (2009) found that management teams that have professional managers with experience outside the company have more proactive strategic orientations. Moreover, ownership and the proportion of founders in the executive team can affect the strategic orientation. Dail and Dalton (1992) could not find any performance differences between founder- and professional-managed firms. Nevertheless, founder departure and the role of founders in the team has been discussed by several scholars (e.g., Virany and Tushman, 1986; Boeker and Karichalil, 2002; Ucbasaran, Lockett et al., 2003; Boeker and Wiltbank, 2005; Chandler, Honig, and Wiklund, 2005). The impact of ownership has been explored in the context of family firms (e.g., Ensley and Pearson, 2005). For example, ownership can affect the power distribution within the decision making group, and power distribution is an important aspect in the decision making process (e.g., Jemison, 1981; Shrivastava and Grant, 1985; Eisenhardt, 1989). In this study, the following research questions are addressed.

Research question 3.4.1: How does the composition of the executive team influence formal strategic planning activities?

Research question 3.4.2: How does the composition of the executive team influence informal strategic planning activities?
Research question 3.4.3: How does the composition of the executive team influence the entrepreneurial strategic posture?

Demographic attributes and general characteristics:

The demographical characteristics of the executives and general attributes of the management team certainly influence the strategy making process. Forbes (2005) pointed out that companies led by older executives tend to make faster decisions. On one hand, age is associated with greater experience and allows executives to make decisions based on informal routines instead of extensive planning and analysis. On the other hand, age has also been related to a more conservative view and less willingness to take risks. Studies looking at executive tenure have argued that managers with a long tenure tend to gather less external information (Miller, 1991) and are less likely to make big organisational changes (Gabarro, 1987). New executives from outside the company are especially more open towards strategic changes and experimentation. Finkelstein and Hambrick (1990, p. 488) noted the following: "Executives with short tenures have fresh, diverse information and are willing to take risks, often departing widely from industry conventions. As tenure increases, perceptions become very restricted and risk taking is avoided. The lowest-risk thing to do is follow the general tendency of mainstream competitors".

In summary, previous studies support the assumption that certain management team characteristics affect strategic decision making. Thus, the following research questions are formulated.

Research question 3.5.1: How do the characteristics of the executive team influence formal strategic planning activities?
Research question 3.5.2: How do the characteristics of the executive team influence informal strategic planning activities?
Research question 3.5.3: How do the characteristics of the executive team influence the entrepreneurial strategic posture?

In table 4.3 the formulated hypotheses and research questions concerning the link between executive characteristics and strategic decision making are displayed.

Table 4.3: Link between executive characteristics and strategic decision making

Prior professional experiences
Hypothesis 3.1.1: Previous professional experience of the executive team affects strategic planning activities.
Hypothesis 3.1.2: Executive teams with prior entrepreneurial, management, or industry work experience emphasise less formal strategic planning activities.
Hypothesis 3.1.3: Executive teams with prior entrepreneurial, management, or industry work experience emphasise more informal strategic planning activities.
Research question 3.1: How does prior entrepreneurial, management, or industry work experience within the executive team influence the entrepreneurial strategic posture?

Educational background
Research question 3.2.1: How does the educational background of the executive team influence formal strategic planning activities?
Research question 3.2.2: How does the educational background of the executive team influence informal strategic planning activities?
Research question 3.2.3: How does the educational background of the executive team influence the entrepreneurial strategic posture?

Heterogeneity
Hypothesis 3.2.1: Heterogeneity in the executive team affects the intensity of strategic planning activities.
Hypothesis 3.2.2: Heterogeneous teams emphasise less formal strategic planning activities than homogeneous teams.
Hypothesis 3.2.3: Heterogeneous teams emphasise more informal strategic planning activities than homogeneous teams.
Research question 3.3: How does heterogeneity within the executive team influence the entrepreneurial strategic posture?

Composition
Research question 3.4.1: How does the composition of the executive team influence formal strategic planning activities?
Research question 3.4.2: How does the composition of the executive team influence informal strategic planning activities?
Research question 3.4.3: How does the composition of the executive team influence the entrepreneurial strategic posture?

Demographic attributes and general characteristics
Research question 3.5.1: How do the characteristics of the executive team influence formal strategic planning activities?
Research question 3.5.2: How do the characteristics of the executive team influence informal strategic planning activities?
Research question 3.5.3: How do the characteristics of the executive team influence the entrepreneurial strategic posture?

4.5   Link between firm and environmental factors and strategic decision making

As discussed in the previous chapters, Carpenter, Geletkancz, and Sanders (2004) have extended the upper echelons framework and included organisational aspects and environmental factors such as uncertainty and dynamism. This study includes academic spin-offs varying in size and age and at different levels in evolutionary development. For this reason, the next paragraph first introduces the concept of the life-cycle phases that companies go through in general. Second, existing models regarding the development of research-based spin-offs are discussed. On this basis, the links between organisational characteristics, environmental uncertainty, and strategic decision making are outlined. Figure 4.5 gives an overview of the relationships examined in this section.

Figure 4.5: Link between firm and environmental level factors and strategic decision making

The objective situation	Upper echelons characteristics	Strategic orientation	Performance
Company maturity: Firm age, Firm size (employees), Life-cycle stage	*Prior professional experiences:* Entrepreneurial experience, Management experience, Industry work experience	*Formal and informal strategy:* Planning, Futurity, Flexibility, Experimentation, Partnering	Growth, Profitability
Environmental uncertainty: State uncertainty, Effect uncertainty	*Educational background:* Management education, Technology education, Level of education	*Entrepreneurial strategic posture:* Innovativeness, Proactiveness, Risk-taking, Competitive aggressiveness, Autonomy	
Technology base: Technology orientation, Research intensity, Product focus	*Heterogeneity:* Diversity experience, Diversity education, Diversity composition		
	Composition: External manager, Ownership, Founder		
	Demographic and general attributes: Team size, Average age, Gender distribution		

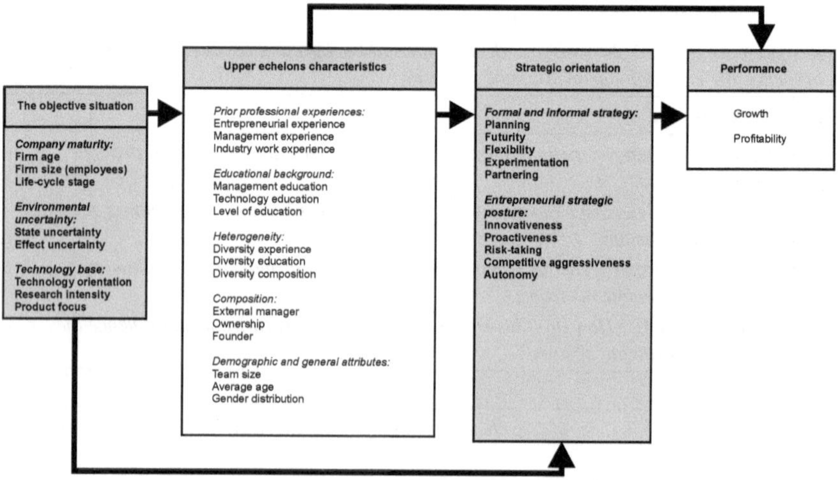

4.5.1   Life-cycle theory and stage-based models

Research in the field of company life-cycles has studied changes in the evolution of firms that develop over a period of time (e.g., Miller and Friesen, 1984; Van de Ven, Hudson, and Schroder, 1984). The success of a new venture depends crucially on the ability of its entrepreneurs to cope with new and different challenges that develop as the business evolves, especially within the scope of research-based high-technology companies. In this context, new ventures can grow from a start-up with a specific marketable product to a complex organisational system in just a few years (Hanks, Watson et al., 1993; Boeker and Karichchalil, 2002). In contrast to more predictable environments, growth-related transitions and crises must be handled earlier and faster (Greiner, 1972).

Life-cycle theory suggests that ventures need significant changes in the managerial capabilities of their leadership teams as the organisations grow and evolve through different life-cycle stages (Hanks, Watson et al., 1993; Covin and Slevin, 1997). Entrepreneurs who have successfully set up an innovative venture may not be perfectly qualified to lead a company through further phases (Galbraith and Vesper, 1982). Because the stages of evolution differ fundamentally from the founding process, entrepreneurs may lack the necessary skills and experience. The main challenge in managing the venture is that it becomes less entrepreneurial, with a focus on viability and survival, and more administrative, with larger organisational systems and even more complex business processes (Tushman and Romanelli, 1985). Life-cycle theorists argue that the continuous involvement of the original entrepreneurs in general management activities may be disadvantageous to the success of a venture as it grows Because the skills of the founding team may not be well suited to manage larger and more established companies, new ventures may need to get professional managers involved in the team to match the required transitions (Adizes 1989). Hambrick and Crozier (1985) empirically observed that successful ventures are more likely to include experienced managers in the growth phase. In contrast, less successful start-ups are more likely to keep the original founding and managing team.

Scholars have used life-cycle and stage-based models to describe the development of new firms. The literature in this field states that organisations face several very similar problems as they grow and age. With respect to this matter, entrepreneurs must change the management and the structural configurations of their organisations considerably (Pugh, Hinkson et al., 1968). According to most life-cycle models, new ventures develop in a coherent and predictable way through different phases of growth. Within each stage of development, differing organisational characteristics imply a number of new challenges and a necessity for different managerial and entrepreneurial qualities (Chandler, 1962; Greiner, 1972; Kazanjian, 1988; Miller and Friesen, 1984). The life cycle models introduced in prior studies relate to each other in their basic idea that the growth of new ventures takes place in predictable patterns and distinguishable time periods, although these models vary in the number of stages. Greiner (1972) proposed one of the first stage-based models and stated that ventures move through five discrete phases. Therefore, firms must make significant changes within their organisations before they can successfully manage the following stage.

Several studies examine the evolutionary process of spin-off ventures and introduce specific models for their particular development path (e.g., Carayannis Rogers et al., 1998; Ndonzuau, Pirnay, and Surlemont, 2002; Roberts and Malone, 1996; Vohora, Wright, and

Lockett, 2004). As in general models, these specific concepts also describe the development process as a number of distinctive phases. Ndonzuau, Pirnay, and Surlemont (2002) outlined four phases: stage (1), when business ideas are generated from research; stage (2), when new venture projects are finalised; stage (3), when spin-off firms are launched from projects; and stage (4), when the creation of economic value by the spin-offs is strengthened. The models introduced by Hindle and Yencken (2004) and Vohora, Wright, and Lockett (2004) are more complex and incorporate a dynamic perspective in that spin-offs may move back and forward through particular stages. Vohora, Wright, and Lockett (2004) described five phases that spin-offs go through: (1) research, (2) opportunity framing, (3) preorganisation, (4) reorientation, and (5) sustainable returns. Additionally, they identified four critical junctures between these different stages as thresholds that the spin-offs must be overcome to reach the next transition phase: (1) opportunity recognition, (2) entrepreneurial commitment, (3) credibility, and (4) sustainability. In contrast to traditional models, spin-off models focus more on early stages. On one hand, they usually do not include a decline phase that describes the evolutionary end of an organisation. On the other hand, models focusing on spin-offs often add an additional phase at the beginning of the venture development, even before the company is officially incorporated, and call it a research phase (e.g., Vohora, Wright, and Lockett, 2004).

4.5.2   Hypotheses and research questions

Mintzberg and Lampel (1999) claimed that firm level strategy evolves when the venture grows, becomes older, or faces different external conditions. This aspect is important in the context of academic spin-offs because, as discussed above, they can change their characteristics significantly. The following sections discuss the potential impact of the maturity of the company, the technology base, and the environmental uncertainty on strategic decision making processes.

Company maturity:

According to the discussed life-cycle models, the maturity of a company is related to the development stage in the evolutionary process. Looking at the strategic decision making process, previous research has observed that, especially in the early phases of venture development, successful entrepreneurs emphasise less formal planning activities (Bhide, 2000). Prior research has looked at aspects such as firm age or size, which both partly reflect maturity. Fredrickson and Iaquinto (1989) observed that changes in the organisational size of a company result in changes in the level of comprehensiveness of strategic decision processes.

The size of a company is generally positively correlated with a better access to resources and a higher level of differentiation and complexity in the organisation, which results in more extensive planning activities and processes (Fredrickson and Mitchell, 1984; Mintzberg, 1973). In line with these findings, Brouthers, Andriessen, and Nicolaes (1998) concluded that executives of small firms tend to rely on intuition in strategic decision making.

According to the previous discussion, the following hypotheses and research question are stated:

Hypothesis 4.1.1: The maturity of the company affects the intensity of strategic planning activities.

Hypothesis 4.1.2: Firms in later development and life-cycle phases emphasise more formal strategic planning activities.

Hypothesis 4.1.3: Firms in early development and life-cycle phases emphasise more informal strategic planning activities.

Research question 4.1: How does the maturity of the company influence the entrepreneurial strategic posture?

Environmental uncertainty:

As outlined in the previous paragraphs, the strategy making process may differ in emerging ventures and established corporations. New companies are confronted with more uncertainty and less predictability because they must build up their businesses without the aid of information about past development or the influence of the environmental situation (McMullen and Shepherd, 2006). Bhide (2004) also connected the lack of comprehensive planning with the uncertain environmental situation. In this vein, new firms are more inclined to enforce learning and apply discovery-driven (McGrath et al., 1995) or logically incremental (Quinn, 1980) approaches than established companies (Shepherd, Zacharakis, and Baron, 2003). With a sample of 130 small business ventures, Matthews and Scott (1995) showed empirically that perceived environmental uncertainty leads to a lower level of strategic planning activities.

Therefore, the following hypotheses and research question address the level of environmental uncertainty and its impact on strategic decision making in academic spin-offs.

Hypothesis 4.2.1: The perceived level of environmental uncertainty affects the intensity of strategic planning activities.

Hypothesis 4.2.2: Firms in an uncertain environment emphasise less formal strategic planning activities.

Hypothesis 4.2.3: Firms in an uncertain environment emphasise more formal strategic planning activities.
Research question 4.2: How does the perceived level of environmental uncertainty influence the entrepreneurial strategic posture?

Technology base:
Another important aspect in the evolution of research-based spin-off companies refers to the technology base. As the specific stage-based models (e.g., Vohora, Wright, and Lockett, 2004) illustrate, spin-offs often start with a research phase. Subsequent to this stage, many ventures first introduce a more service-oriented portfolio to generate revenues (Leitch and Harrison, 2005) so that financial returns can be used to develop the technology and products. Thus, depending on the development stage, spin-offs may be more service- or more product-oriented. Furthermore, spin-offs may rely on extensive technology and research to varying degrees. In this context, Roberts (1991b) studied the technology base of new ventures by referring to a sample of 125 university spin-offs and defined four categories that classify the importance of the underlying technology: direct, partial, vague, and none. Borch, Huse, and Senneseth, (1999) examined the relationship between the resource configuration and strategic orientation in small companies. Based on a dataset of 660 firms, they concluded that technological firms apply growth and product strategies to a greater extent.

Because empirical evidence links the technology base and strategic decision making, the following research questions are stated.

Research question 4.3.1: How does the technology base influence the level of formal strategic planning activities?
Research question 4.3.2: How does the technology base influence the level of informal strategic planning activities?
Research question 4.3.3: How does the technology base influence the entrepreneurial strategic posture?

Table 4.4 shows all hypotheses and research questions regarding the examined relationships between firm and environmental factors and strategic decision making.

Table 4.4: Link between firm and environmental factors and strategic decision making

Company maturity
Hypothesis 4.1.1: The maturity of the company affects the intensity of strategic planning activities.
Hypothesis 4.1.2: Firms in later development and life-cycle phases emphasise more formal strategic planning activities.
Hypothesis 4.1.3: Firms in early development and life-cycle phases emphasise more informal strategic planning activities.
Research question 4.1: How does the maturity of the company influence the entrepreneurial strategic posture?
Environmental uncertainty
Hypothesis 4.2.1: The perceived level of environmental uncertainty affects the intensity of strategic planning activities.
Hypothesis 4.2.2: Firms in an uncertain environment emphasise less formal strategic planning activities.
Hypothesis 4.2.3: Firms in an uncertain environment emphasise more formal strategic planning activities.
Research question 4.2: How does the perceived level of environmental uncertainty influence the entrepreneurial strategic posture?
Technology base
Research question 4.3.1: How does the technology base influence the level of formal strategic planning activities?
Research question 4.3.2: How does the technology base influence the level of informal strategic planning activities?
Research question 4.3.3: How does the technology base influence the entrepreneurial strategic posture?

5 Preliminary study and main survey

The following chapter describes the data collection procedure for the empirical analysis and is divided into two main parts. First, a preliminary study was conducted to characterise the entire spin-off population because none of the existing and accessible external sources provide such a comprehensive overview. This survey requested that technology transfer officers and directors of research organisations in Europe provide general information regarding the companies that had spun off from their institutes. Second, the main study, which consists of a survey that was sent to managing directors of a selected sample of research-based spin-off companies in Germany, is introduced. This section provides an overview of the sample selection and the development of the survey instrument as well as a descriptive analysis of the group of respondents in comparison with the sample and the population.

5.1 Preliminary study and population

The preliminary study addressed research-based spin-off companies in Europe. ProTon Europe, a European Knowledge Transfer Association, collected data pertaining to academic spin-offs in 2005, but the database is not accessible for non-members and is no longer updated. In fact, there is no existing public database of spin-offs in Europe that can describe the main characteristics of the entire population. The need to understand the overall spin-off population marked the starting point for the preliminary study.

5.1.1 Data collection

In cooperation with the University of Antwerp (Professor Dr. Johan Braet and Dr. Sven De Cleyn), a preliminary study was conducted to investigate the initial questions regarding how many spin-off ventures originated from universities and non-university research organisations during the period from 1985 to 2008 and how these companies can be characterised. Intensive research that was found on websites and databases yielded information pertaining to 1,011 academic organisations and contact details of potential respondents at these organisations. Ideally, the respondents included individuals who were employed in the technology transfer office or a similar department that is responsible for the commercialisation and spin-off processes. In situations in which an organisation did not have a specific contact person or office for these issues, the general information address was used.

In summary, 809 academic organisations in 24 European countries were contacted via an email survey that was composed in English and translated into German, French, and Dutch to increase response rates. The survey requested the names and websites of the spin-offs of the organisations and their corresponding dates of foundation, activity statuses, academic fields, and industry sectors as well as information regarding the involvement of their parent organisations, such as equity stakes, patent licences, or infrastructure. The survey was followed by two reminders, so all these activities led to a final response rate of 40.3 per cent. In addition, some research institutes did not respond to the requests for information but provided information regarding their spin-offs on their websites. In total, 8,507 spin-off companies from 384 organisations in 24 European countries were identified; thus, it was possible to collect data on spin-offs from 47.6 per cent of the contacted public research organisations in Europe. However, respondents frequently delivered incomplete information; for example, many respondents returned only a list of the names of the spin-offs without any further details. Therefore, during the following period of time, intensive research efforts were undertaken to gather additional information pertaining to these spin-off companies. Nevertheless, most of the descriptive analyses in the following section refer to only a subsample rather than to the entire population.

A total of 86 per cent of the spin-offs in the sample are still active or established, whereas 10 per cent are inactive, have ceased operations, or have failed. Approximately 4 per cent of the ventures have already merged with other entities or have been acquired or sold. A high percentage of the spin-offs were founded in technology-based industries (72 per cent). The following sections provide a brief statistical overview of the results with regard to the population of research-based spin-off companies in Europe.

5.1.2   Descriptive overview

The statistical overview is guided by the main purpose of the preliminary study: to identify the population of spin-off companies in Europe and to describe some of their key characteristics. First, this section provides an overview of the number of identified spin-off companies in 24 European countries and the average number of spin-offs per academic organisation. In this section, the distribution of the years of foundation of spin-offs in a selected subset of these countries, including Belgium, France, Germany, Italy, the Netherlands, Spain, Sweden, Switzerland, and the UK, is also briefly discussed. The second part contains an examination of industry sectors and compares the distributions of technology-oriented and non-technology-oriented companies. This part examines technology-oriented industry sectors in greater detail and compares the findings of this preliminary study to the results of other empirical studies of high-technology ventures in Germany. Third, the size of the management teams and the number of shareholders of the spin-off ventures are shown and related to other external findings. In the fourth section, the information regarding both the turnover and the number of employees provides insights into the economic figures of academic spin-offs. Each paragraph briefly describes the data collection procedure and outlines the main results and their limitations. Subsequently, the findings are compared with those of other studies in the field.

Table 5.1: Overview of identified spin-offs per country

No.	Country	Number of spin-offs	Proportion	Number of organisations	Average number of spin-offs per organisation
1	Austria	290	3.41%	12	24.2
2	Belgium	417	4.90%	20	20.9
3	Czech Republic	11	0.13%	6	1.8
4	Denmark	107	1.26%	2	53.5
5	Finland	199	2.34%	8	24.9
6	France	1,016	11.94%	50	20.3
7	Germany	2,071	24.34%	96	21.6
8	Greece	88	1.03%	4	22.0
9	Iceland	85	1.00%	2	42.5
10	Ireland	32	0.38%	3	10.7
11	Italy	797	9.37%	41	19.4
12	Liechtenstein	0	0.00%	1	0.0
13	Luxembourg	2	0.02%	1	2.0
14	Netherlands	646	7.59%	9	71.8
15	Norway	11	0.13%	1	11.0
16	Poland	38	0.45%	3	12.7
17	Portugal	233	2.74%	12	19.4
18	Serbia	43	0.51%	2	21.5
19	Slovakia	2	0.02%	2	1.0
20	Slovenia	20	0.24%	1	20.0
21	Spain	410	4.82%	23	17.8
22	Sweden	542	6.37%	19	28.5
23	Switzerland	504	5.92%	12	42.0
24	United Kingdom	943	11.08%	54	17.5
	Total	**8,507**	**100.00 %**	**384**	**22.2**

Table 5.1 presents the number of spin-off companies that were identified within the preliminary study. This table provides details regarding the number of spin-offs per country and the average number of academic organisations from which they were derived. A total of 8,507 spin-offs in 24 different European countries were found. Most of these spin-offs were identified in Germany (2,071), but other countries with a considerable number of spin-offs include Belgium (417), France (1,016), Italy (797), the Netherlands (646), Sweden (542), Switzerland (504), and the United Kingdom (943). The current status of the companies could be ascertained only for a subset of the spin-off population. Figure 5.1 shows that 86 per cent of the spin-offs in this subsample are still active or established, whereas 10 per cent of the spin-offs are inactive, have ceased operations, or have failed. Approximately 4 per cent of the ventures have already merged or have been acquired or sold.

Figure 5.1: Current status of the spin-offs

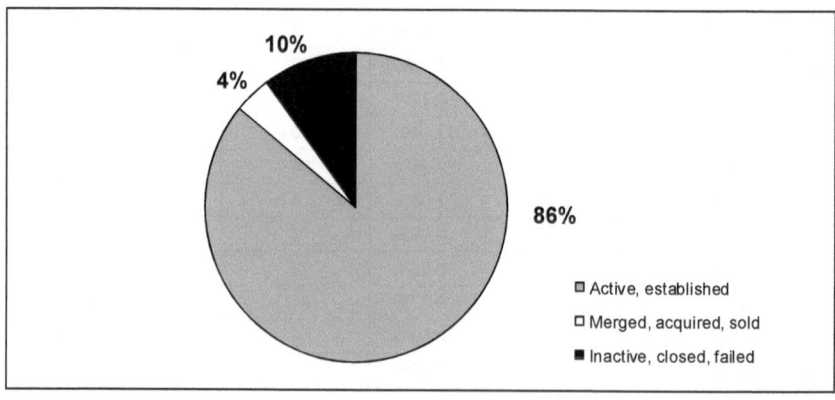

N = 3,115

The results must be linked to other studies in this domain to demonstrate how the sample reflects the entire spin-off population in Europe. In a study by the Centre for European Economic Research, Egeln, Gottschalk et al. (2002) conducted a survey that was designed to examine spin-off foundations from publicly financed research institutes in Germany from 1996 to 2000. Rather than approaching the managers of technology transfer offices or the directors of research institutes, these authors collected a stratified random sample of the *ZEW-Gründungspanel*, which is a database that tracks newly founded companies in different industry sectors in Germany. The results of the survey, which included 20,000 companies, were utilised as a basis to extrapolate the entire spin-off population in the period of time that was considered. According to the underlying definition of their study, these authors calculated an average number of 6,800 new spin-offs per year in Germany. However, the study defined two different types of spin-off companies. The *Verwertungs-Spin-off* category includes companies that were founded on the basis of the exploitation of research findings, and the definition that was used in this categorisation is similar to the definition that is used in this study. The authors estimated that approximately 2,600 new spin-offs per year belong to this category. The second category was labelled *Kompetenz-Spin-off* and used a broader definition that also includes spin-offs based on individual knowledge.

In the data analyses, the number of identified spin-offs likely does not reflect the level of spin-off activity in each country. There are several possible reasons for these limitations. One cause of these limitations could be the data collection procedure. Presumably, researchers have better access to research organisations and their contact persons in those countries in which they are working (i.e., Germany and Belgium) because of their social networks within these countries. This close access could cause these countries to be overrepresented. Second, in this step, the quality of the received data in comparison with the quality of the requested data could not be demonstrated. Furthermore, only a limited number of failures and mergers or acquisitions could be identified. Technology transfer officers and directors of research institutes might not be able to provide a complete overview of all of the spin-off activities at their organisations and to track their development over time. Therefore, companies that were established long ago and ventures that had already closed or had been sold to other companies are likely to be underrepresented in the sample.

In the next step, the evolution of spin-off activities is discussed. Figure 5.2 shows the number of spin-offs that were founded per year in a subsample of the study in the period from 1985 to 2008. The subset contains countries with a substantial number of spin-offs, such as Belgium, France, Germany, Italy, the Netherlands, Spain, Sweden, Switzerland, and the UK, These countries were also selected with the aim of including different regions of Europe, such as Scandinavia, Benelux, Western Europe, Central Europe, and Southern Europe. The distribution that is shown in figure 5.2 discloses an overall increasing level of spin-off activity from 1985 to 2008 with a peak in 2003 followed by a decreasing trend in new venture formations from 2004 to 2008. Furthermore, several differences between the selected countries can be outlined. For example, according to these data, the spin-off phenomenon began later in Italy, Spain, and France than in the other countries. In addition, in most years, a fairly constant number of new spin-offs per year can be observed in Switzerland, the Netherlands and Belgium, whereas the other countries show remarkably increasing levels of activity until 2003 and decreasing levels from 2004 to 2008.

The data might lack information pertaining to early and more recent spin-offs for the following reasons. As mentioned above, managers of technology transfer offices and directors of research institutes might have only limited access to information regarding spin-offs that were established long ago because many technology transfer offices were created rather recently. Some of the information was obtained from different sources; in addition to survey responses, organisational websites were used.

Figure 5.2: Number of spin-offs founded per year

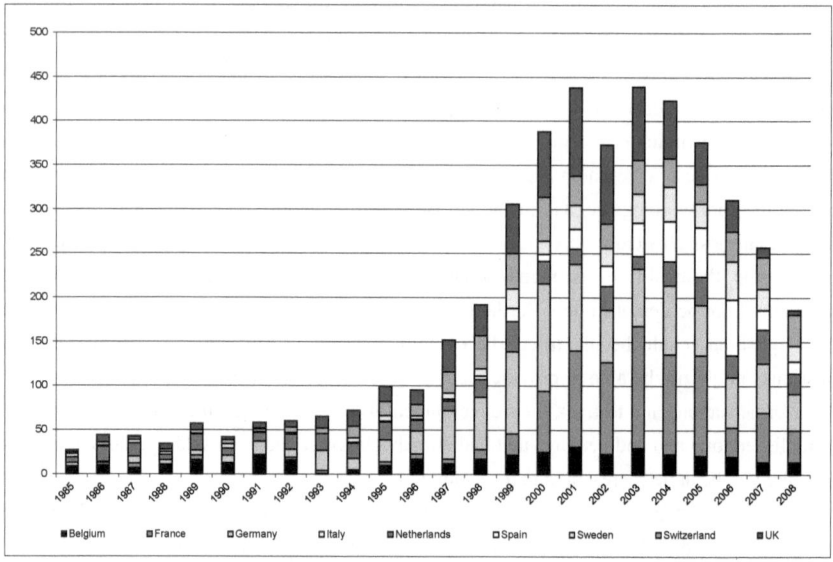

N = 6,066

A comparison of these results to those of other studies reveals that universities and research institutes have recently become increasingly entrepreneurial. Etzkowitz (1998) coined the term *entrepreneurial university* to refer to this phenomenon. Several initiatives and support schemes from executives on both the national and the European levels attempt to encourage the transfer of publicly financed research findings to the business environment. Another consequence of the entrepreneurial university is a rapidly growing body of literature in the field of academic entrepreneurship. Figure 5.3 shows the number of articles that were published in regular and special issues per year from 1981 to 2005 (Rothaermel, Agnung, and Jiang, 2007). These data demonstrate that the research in this field increased significantly in the late 1990s. Therefore, these observations show two sides of the same issue.

Figure 5.3: University entrepreneurship articles published per year

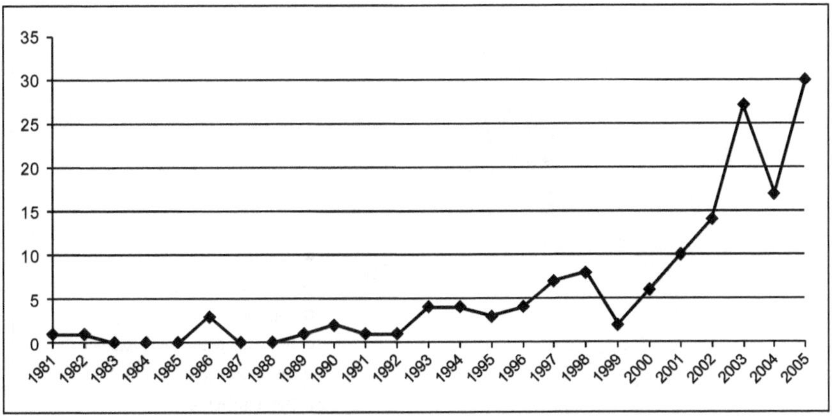

Source: Rothaermel, Agung, and Jiang (2007, p. 696)

The first step involves obtaining an overview of the distribution of industry sectors within the population; in this step, a subsample of German spin-offs was selected, and the main industry sector of each was identified by using company databases and cross-checking the outcomes with the content of company websites. The spin-offs belonged to more than 60 different NACE industry sectors; thus, the spin-offs were clustered into groups to make the results more accessible and comparable to other studies. The classification system follows studies by the Fraunhofer-Gesellschaft and the Centre for Economic Research in Germany (e.g., Egeln, Gottschalk et al., 2002; Hemer, Walter et al., 2005) and contains the following groups: (1) high technology, (2) advanced technology, (3) software, (4) technology-based services, (5) other services, and (6) other industries. Figure 5.4 provides an overview of the main industry sectors of spin-off companies in Germany.

A high percentage of spin-offs were founded in technology-based industries (72 per cent), including the categories of high technology, advanced technology, software, and technology-based services. The high technology cluster contains industry sectors that include manufacturing companies, which spend an average of more than 8 per cent of their total turnover on research and development. By contrast, the advanced technology category includes manufacturing companies with a research and development stake of between 3.5 per cent and 8 per cent (Grupp, Jungmittag et al., 2000). More than half of the spin-off companies belong to the service sector (54 per cent). The technology-based services category follows the classification of the Federal Statistical Office in Germany.

Figure 5.4: Industry sectors of German spin-off companies

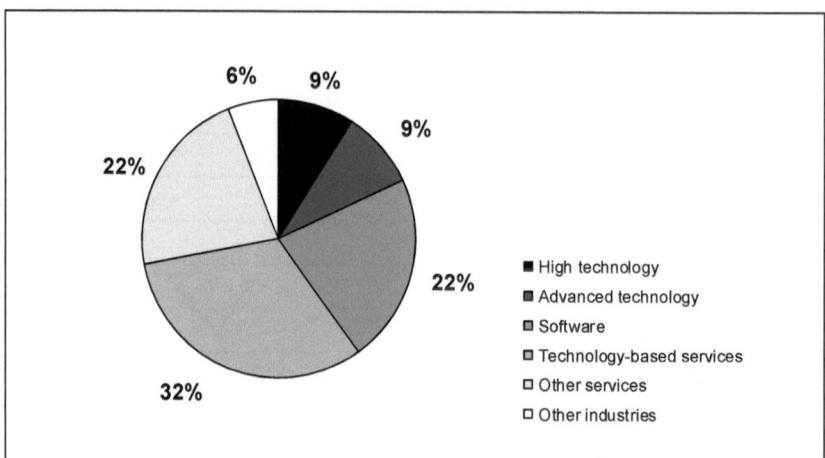

N = 748

The distribution of industry sectors in this study must be compared to that of other studies in this field to draw conclusions from the findings. The *Gründungspanel* is an annual report of the Centre for European Economic Research and the KfW Bankengruppe that analyses company formations across different industry sectors in Germany and tracks their development over a certain period of time. Figure 5.5 shows that less than 10 per cent of company foundations in the Gründungspanel, which does not focus on any specific sectors, occurred in technology-based industries. This result supports the conclusion that spin-offs account for only a narrow subgroup of newly formed companies, whose special characteristics must presumably be analysed and discussed separately.

Figure 5.5: Newly formed companies in technology and non-technology industries

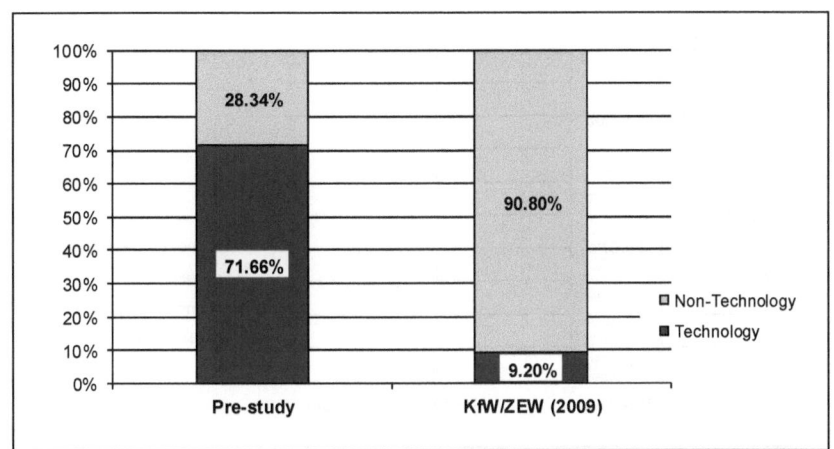

In their study "High-Tech Gründungen in Deutschland", Metzger, Niefert, and Licht (2008) focused on companies that were founded in technology-oriented industries in Germany from 1995 to 2007. The report is a continuation of former studies and was published by the Centre for European Economic Research. Because these authors used the same classifications as those that are used in the present study, the distribution of industry sectors within the samples can be compared. Figure 5.6 illustrates the distribution of the technology-based industry sectors in the preliminary study and compares this distribution with that of Metzger, Niefert, and Licht (2008). Although there are several differences between the studies, their overall patterns of results are similar. In both studies, the technology-based service sector has the highest percentage of companies, and the proportion of ventures in the software sector is similar in both studies (approximately 30 per cent). The high technology and advanced technology categories follow in third and fourth places, respectively, and both categories are similar in size (approximately 12.5 per cent and 8 per cent, respectively).

All of the samples that are discussed are derived from German companies. Therefore, it might be problematic to offer general propositions with regard to the industry sectors of spin-offs in other European countries. In addition, the small size of the subsample that was considered within the preliminary study might cause difficulties in discussions of its differences from other studies.

Figure 5.6: Newly formed companies in technology-based industries

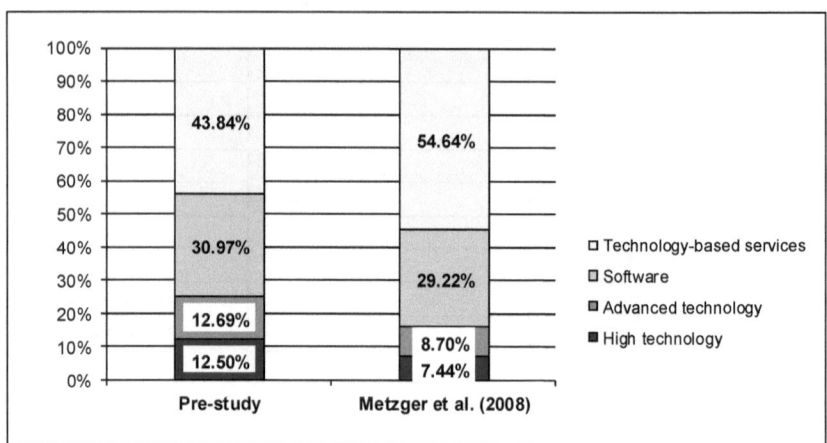

Figure 5.7 shows the sizes of the management teams across the technology-based sectors. The high technology and advanced technology sectors have the lowest proportions of companies whose management teams contain only one member. The maximum size is 6 managers. The typical spin-off management team consists of one or two managers. The average number of managers can be described by the overall mean, which is 1.7, and the overall median, which is 2.0. Several studies have demonstrated a correlation between the type of spin-off and the number of founders (e.g., Egeln, Gottschalk et al., 2002). The main argument of these studies is that technology-based ventures require a range of competences (human capital) that is broader than that of non-technology-oriented companies; therefore, technology-based ventures are more likely to be founded by teams rather than by single individuals (Roberts, 1991a). These figures show only the size of the management teams and do not display the number of founders.

Figure 5.7: Size of the management teams

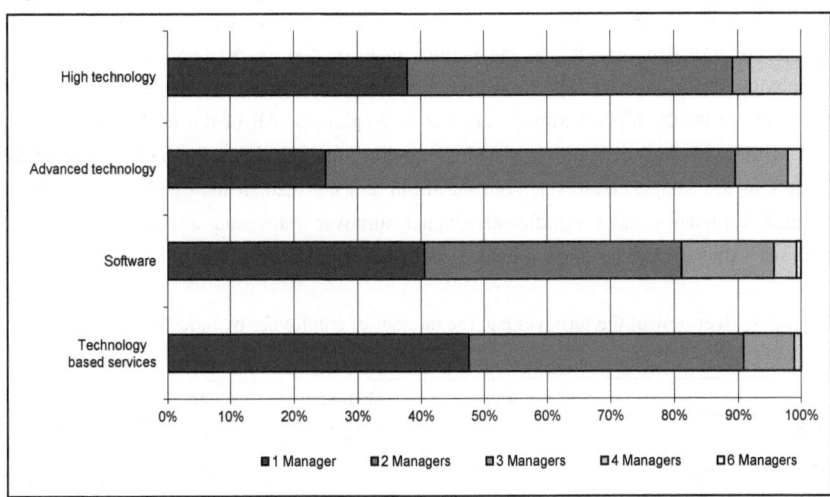

N = 408

Table 5.2 shows the average number of shareholders of the spin-offs in the technology-oriented industry sectors. The overall average, which is represented by the mean value of 4.3 and the median of 3.0, is higher than the average number of managers. The high-technology sector has the highest average, which is indicated by both the mean and the median. The maximum number of shareholders is 29 and includes several stakeholders, such as venture capitalists.

Table 5.2: Number of shareholders

	Mean	Median	Minimum	Maximum
Total	4.3	3.0	1.0	29.0
High technology	6.2	4.0	1.0	29.0
Advanced technology	3.7	3.0	1.0	16.0
Software	3.7	3.0	1.0	23.0
Technology-based services	5.2	3.0	1.0	23.0

N = 511

Table 5.3 provides an overview of the turnover and number of employees of the spin-off companies and displays both the overall average and the median are displayed. All of these figures are provided for each of the various industry sectors. Ventures in the advanced technology sector have the highest average turnover, whereas the high-technology sector is characterised by the highest average number of employees. All of the figures demonstrate large differences between the mean and the median; such differences might be caused by outliers in the sample. However, both the mean and the median indicate that technology-oriented companies have significantly higher turnover rates and a greater number of employees than companies in other industry sectors.

Table 5.3: Overview of the turnover and the number of employees by industry

	Turnover in EUR (N = 383)		Employees in FTE (N = 646)	
	Mean	Median	Mean	Median
Total	2,142,869	900,000	16.6	7.0
High technology	3,045,295	1,500,000	28.1	11.0
Advanced technology	4,021,250	1,275,000	26.3	10.5
Software	2,559,925	1,006,000	19.0	8.0
Technology-based services	2,014,676	800,000	17.2	7.0
Others	1,015,950	500,000	13.3	5.0
Services	989,931	560,000	9.4	5.0

Because the median is generally more robust to outliers, the two measures would probably be more similar in a larger sample. Figures 5.8 and 5.9 compare the median of the turnover and the number of employees with the means for the population in the study that was conducted by Metzger, Niefert, and Licht (2008). Interestingly, a large difference can be found only for the turnover in the software sector.

Figure 5.8: Turnover values by industry

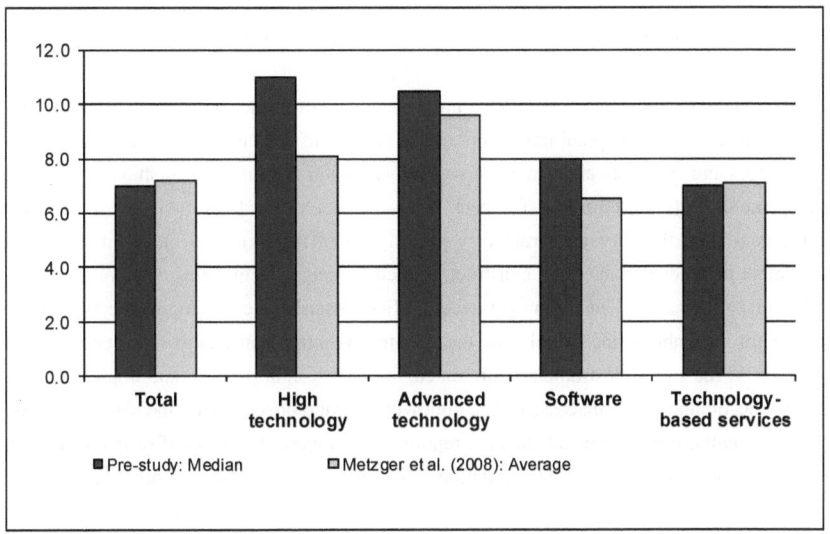

Figure 5.9: Number of employees by industry

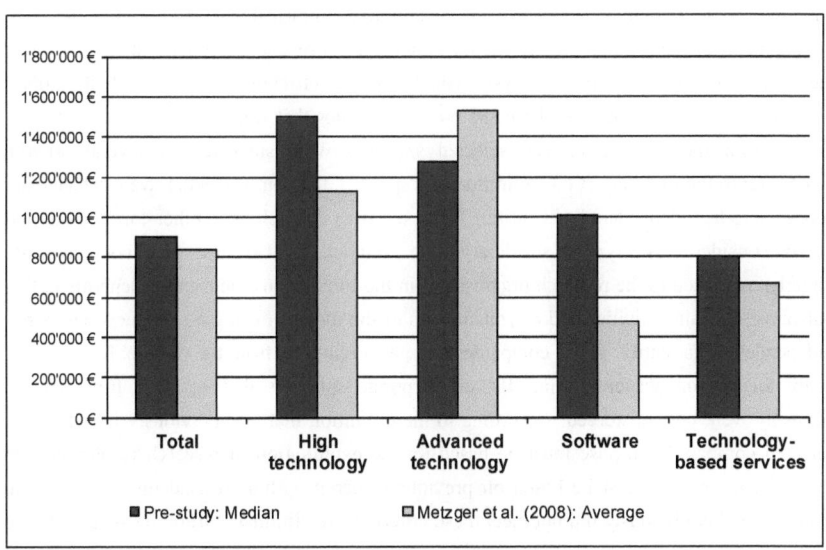

5.2   Main survey and data collection

This study aims to provide support for the hypotheses and to examine the research questions that are outlined in the previous chapter. These questions address the executive team characteristics, strategic decision making processes, and performance of research-based spin-off companies. The conceptual framework of this study was presented at several international conferences and symposia and discussed with experts from the entrepreneurship and strategic management fields. Based on the feedback that was received, multiple revisions were made. In the next step, the survey instrument was designed in the form of a questionnaire. This instrument includes variables and constructs related to management teams, venture strategies, and firm performance. The following sections first describe the selection process and the criteria for assembling the sample for the quantitative empirical analysis. Second, a brief overview of the survey instrument is presented, and the design, pre-test, and implementation of the questionnaire are discussed. The following paragraph describes the data collection process. Finally, descriptive statistics are reported to compare the group of respondents with the selected sample and the initial population and to discuss similarities and differences.

5.2.1   Sample selection

An appropriate sample is necessary for quantitative empirical analysis. A high level of homogeneity in the sample reduces the unwanted effects of unobserved variables that lower the quality of further causal analyses (Backhaus, Blechschmidt, and Eisenbeiß, 2006). Therefore, a sample of research-based spin-off companies that were located in Germany was drawn from the population. The selected spin-offs were subjected to several criteria. According to the first part of the definition of a spin-off, the companies that were selected are supposed to be independent legal entities. Thus, all firms that belong to other corporations and can be considered as a type of subsidiary were excluded. Furthermore, some companies are directly connected to the research organisation in the form of an extension or department that outsources certain activities and responsibilities of the mother institution to a new and partly independent legal entity. These companies were also removed from the sample. Because this study focuses on academic spin-offs, all corporate spin-offs that emerged from another company were not considered. According to the definition that was previously introduced, a selected company's purpose must be based on the exploitation of research results, and its initial business idea must be based on previous research within an academic organisation. Companies that obviously did not meet these criteria were eliminated from the sample. In the next selection step, only corporate entities were considered because of the difficulty that is associated with gathering information regarding private companies from external data sources. In addition, private companies are rare among research-based spin-offs because they usually do not provide an adequate legal framework to exploit intellectual property. Non-profit organisations were not considered for the sample as well. Firms were also excluded if there was no further information available that could positively confirm that they actually exist, for example, via their own corporate websites or via external databases. The outlined selection process resulted in a sample of 904 companies that were located in Germany. There were no limitations in terms of the industry sectors that were represented in the data. In the next step, detailed information pertaining to all companies was compiled through corporate websites and external sources (e.g., database Amadeus), and additional information was included in the spin-off datasets (e.g., names of managing directors, contact details).

5.2.2 Survey instrument

The quantitative information for this cross-sectional study was collected with an email-based, self-administered survey. The questionnaire was distributed in PDF format and could be completed electronically and returned via email, fax, or postal mail. In comparison with the collection of data via personal interviews, the electronic survey strategy has some advantages. The primary advantage of this survey type is objectivity because respondents are not influenced by interviewers (Bauer and Wölfer, 2001). Therefore, this type of survey is free of interviewer bias. Furthermore, respondents can answer questions at their own convenience and are probably able to reflect on their responses or to seek additional information if necessary. From the perspective of respondents, anonymity is another key advantage because it increases the likelihood of participation and ensures that honest answers are provided (Dommeyer and Eleanor, 2000). In this survey, the responses could be returned anonymously, but only two questionnaires were returned via postal mail or fax and did not include any information regarding the participant.

Nevertheless, several disadvantages of mail surveys must be mentioned. The data collection process is uncontrollable in some ways. Thus, there is no assurance that the respondents have fully understood the questions, and it may be difficult for respondents to request clarification. During the data collection process, only three participants actively asked for clarification regarding the meaning or purpose of certain questions in the survey. In addition, it is theoretically possible that the managing directors who were supposed to answer the questions did not complete the questionnaires themselves. They might have delegated the task to other individuals within their companies who were not considered to be key informants for the underlying questions. On average, mail surveys generate lower response rates than individual interviews; thus, issues regarding representativeness need to be discussed in the analysis.

The design of the questionnaire is also important because the design could influence the outcome and the quality of the data that were collected. The questions are structured and standardised to reduce bias in the answers. The questionnaire for this study is divided into five sections, which are visible to the respondents. As recommended by Schumann (2000), the most easily answered questions, such as those requesting basic information regarding the respondents' companies, are placed at the beginning of the survey, whereas sensitive topics, such as turnover figures, are placed at the end of the questionnaire. The time that is necessary to complete the survey is also crucial. If respondents are given a greater amount of time to complete the survey, then they will be more likely to cease participating halfway through the questionnaire or to not begin the survey at all. For this study, respondents needed approximately 15 minutes to answer all of the questions. The first part of the questionnaire explained practicalities of the survey and provided essential definitions. In addition, this part outlined the purpose and the target group of the study and emphasised that the survey is a key part of a research project that is being conducted at the university, which does not have any commercial interests. The introductory part of the survey also guaranteed the anonymity of the respondents and the non-disclosure of the information that they provide. Finally, this section offered several ways that the participants could provide feedback and gain incentives for their participation; such incentives may include the provision of a summary of the final study or an individual analysis of a respondent's company.

The questions in the second part of the survey requested general information regarding the companies, such as their parent institutions, founding teams, and industry sectors. This part also asked respondents to specify their positions within their companies to ensure that the intended respondents (key informants) were involved. In the third part, the respondents were instructed to provide information regarding the characteristics of their company's current management team. This information included general attributes of a company, such as the size and age of its management team; aspects that reflect the team's composition, such as ownership or the number of external managers; and team characteristics, such as prior professional experience or educational backgrounds. The fourth part of the survey involved several aspects related to the strategic orientation of the spin-off companies. The entrepreneurial orientation construct, including the dimensions of innovativeness, proactiveness, risk-taking, competitive aggressiveness, and autonomy, was measured. This section also sought information regarding the intensity of formal planning activities at the company level because such information reflects the rationality of a company's approach to its decision making processes. Other questions were posed with regard to heuristic and informal strategy processes, such as flexibility or experimentation. In the final part of the survey, respondents were asked to provide estimates regarding the environmental situation and performance of their companies.

Before the data collection process began, intensive and systematic pre-tests were necessary (Hunt, Sparkman, and Wilcox, 1982; Prüfer and Rexroth, 1996). The questionnaire was tested with respect to the comprehensiveness and consistency of the questions to reduce measurement errors and to improve the quality of the data (Diekmann, 2007; Schnell, Hill, and Esser, 2008). To achieve this goal, the first version of the questionnaire was discussed with experts and researchers in the entrepreneurship or strategic management field who possess a background in quantitative empirical studies. The feedback that was received resulted in a major revision of the questionnaire. The next step included two separate pre-tests with 16 spin-off companies in Germany and Belgium. The respondents were observed while they completed the survey and were interviewed after they finished it. The practical feedback led to several modifications of questions that needed further clarification and of the intended procedure to improve the response rate. Finally, the technical functionality of the survey was tested. The final version of the questionnaire is displayed in the appendix.

5.2.3 Data collection process

The survey data were collected between May 2010 and August 2010. In the name of the Chair for Entrepreneurship and Innovation at the University of Technology Dresden, the survey was distributed in two waves. The first wave targeted 425 research-based spin-off companies between May and July. The second wave, which began in June and ended in August, encompassed 479 companies. The 6-pages PDF questionnaire was attached to an email that was addressed to all managing directors of each spin-off. During the process of preparing the survey, the email addresses of all members of each management team were collected in an intensive research effort. Thus, more than 80 per cent of the emails were sent to the personal accounts of the managing directors. If this information was not available for some companies, then the email was sent to the general email address for these companies.

The respondents were instructed to complete the questionnaires electronically or manually and return them via email, fax, or postal mail. If the spin-off companies did not respond in any way (even to state that they were not interested in participating), then two reminders that included the questionnaire were emailed. The follow-up email also asked whether the firms met the definition of a research-based spin-off company and belonged to the target group of the study. Some emails could not be delivered. Several companies had already been acquired, merged, or sold. A considerable number of respondents claimed that their companies did not belong to the target group because their companies were not spin-offs of academic institutions. Finally, all of the companies that did not respond or provide feedback even after the second reminder were contacted by telephone. During the telephone calls, these companies were asked whether they belonged to the target group and whether they were interested in participating in the study. Several incentives were offered to increase the overall response rate. The participating companies were offered the report on the preliminary study, the final report of the main study, and an individual analysis of their companies compared with the entire sample. For companies that participated in the survey, one member of their management teams was contacted to seek clarification regarding inconsistencies and to validate the information that was received. These individuals were asked to complete a short version of the survey to obtain multiple answers.

5.2.4   Overview of the sample

Table 5.4 provides an overview of the outcome of the data collection process. A total of 294 companies from the original sample of 904 were excluded for several reasons: the email could not be delivered; the firms declared that they were not research-based spin-offs belonging to the target group; the firms were acquired, merged or sold; or the firms had failed, declared bankruptcy, or ceased operations. In fact, a high percentage of the firms in the original sample are not spin-offs according to the definition in this study because they do not exploit any type of research results or because they were actually established as corporate spin-offs. Thus, as discussed in chapter 2, accurately targeting academic spin-offs is highly problematic. The final sample consisted of 610 academic spin-offs in Germany. In total, responses from 200 ventures were received: 88 responses in the first wave and 112 responses in the second wave. Of these responses, 7 responses contained empty, damaged, or incomplete files and had to be excluded from further analyses. The final data set comprised 193 companies; thus, the effective response rate was 31.64 per cent.

Table 5.4: Overview of the sample selection

	Number	Percentage	
Contacted (original sample)	904	100.00%	
Undelivered, not sent, deleted	70	7.74%	
Acquired, merged, sold	17	1.88%	
Bankrupt, closed, failed	2	0.22%	
Not a spin-off	205	22.68%	
New sample	**610**	**67.48%**	**100.00%**
Responses	200		32.79%
Empty files, error, incomplete	7		1.15%
Final data set	193		31.64%

5.2.5 Representativeness

A sample is considered to be representative if its composition is consistent with that of the entire population (Bortz and Döring, 2006). Tests for representativeness must demonstrate whether there are systematic differences between the respondents and the sample and whether the composition of the sample is consistent with that of the entire population. As mentioned in the second chapter, no comprehensive overview of the entire spin-off population is available. The preliminary study was designed to gain a better understanding of the spin-off population in Europe. However, many spin-offs were eliminated from the database during the sample selection process because they did not meet the underlying definition of an research-based spin-off. When the non-respondents were contacted by telephone to remind them about the survey and to request feedback, many of these individuals stated that their companies do not view themselves as spin-offs. Different reasons for this viewpoint were given, but the main reason was that the firms were not based on the transfer of knowledge or technology. Because the characteristics of the entire population depend on the underlying definition of an academic spin-off, it is almost impossible to create a sample that is representative of all spin-offs. This aspect was already discussed in the second chapter, which introduced different definitions of academic spin-offs. Nevertheless, the following section compares the data from the present study with the data that were gathered in the preliminary study. The foundation years (reflecting their age), geographical locations, and industry sectors of the companies are discussed.

Figure 5.10 shows that there are differences in the founding year among the population that was drawn from the preliminary study, the selected sample, and the respondents. The years of firm incorporation were clustered in five groups to increase the visibility of the distributions. The main differences between the sample and the population indicate that more companies in the sample were established in the time frame between 1995 and 1999, but fewer companies were established during the period from 2005 to 2008. There were relatively fewer firm foundations in the group of respondents than in the population between 2000 and 2004, but more firms were founded during the period from 1995 to 1999. The distributions of the sample and the respondents differ primarily for the periods between 2000 and 2004 and between 2005 and 2008.

Figure 5.10: Foundation years of pre-study, sample, and responses

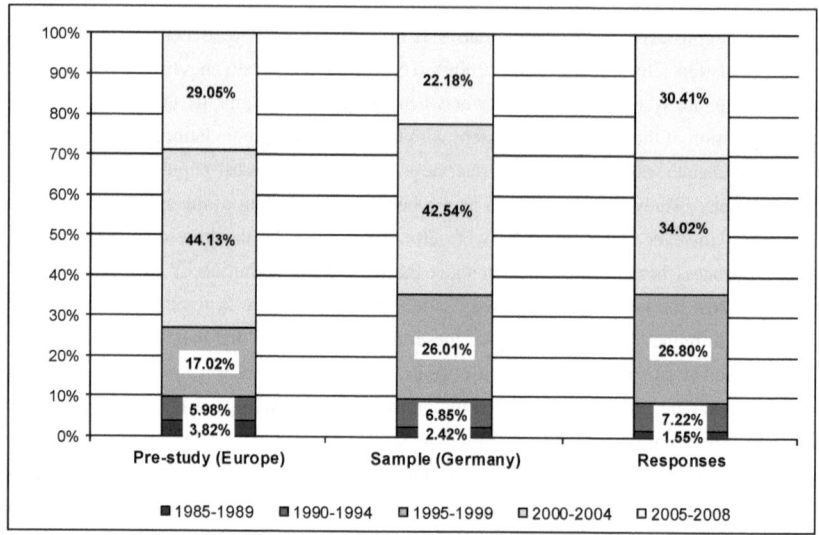

In figure 5.11, the geographical locations of the respondents are compared with the distributions in the sample and in the population. The data for the population are based on the information that was collected in the preliminary survey. The first number of the postal code was used for the formation of different groups. There are only slight differences between the companies in the population and those in the selected sample. However, the figure shows that there exist differences between the group of respondents and both the sample and the population. In the respondent group, more companies have the first digit of 0, which corresponds to the area in Eastern Germany where the University of Technology Dresden is located. The proximity to the university and, thus, likely a greater willingness to support the study might be the reasons for the higher response rate. Spin-offs in this group of postal codes account for 10.05 per cent of the respondents but represent only 5.88 per cent of the population and 6.43 per cent of the sample. Further visible differences can be observed in postal codes that begin with 4 and 5, which are overrepresented in the group of respondents compared with the sample and the population. These two areas are located in Western Germany. In most of the other areas, there are proportionally fewer firms in the group of respondents than in the population or in the sample, but the differences are smaller.

Figure 5.11: Geographical location of pre-study, sample, and responses

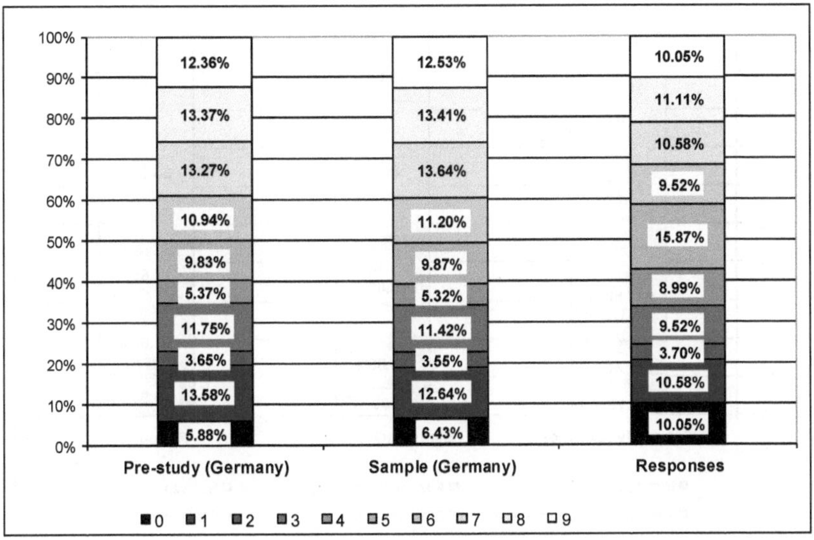

Figure 5.12 displays differences according to various industry sectors. As described in the previous sections, the data for the population are drawn from the preliminary study. There are small differences between the population and the sample, and these differences are most visible in the *Other services* group. In the population, 18.29 per cent of companies belong to this category, whereas only 13.86 per cent of companies in the sample belong to this group. This result indicates that a selection bias was present in the creation of the sample and that this bias reduced the likelihood that this type of company would be included in the analysis. This bias affected the other categories to a lesser extent. In the group of respondents, high technology and advanced technology firms are overrepresented and account for 10.92 and 17.65 per cent of the companies, respectively. In the population, only 6.19 per cent of the companies belong to the high technology category, and 12.83 of the companies belong to the advanced technology category. However, the respondent group contains fewer software firms (17.65 per cent) compared with 24.63 per cent in the population and 26.00 per cent in the sample.

Figure 5.12: Industry sectors of pre-study, sample, and responses

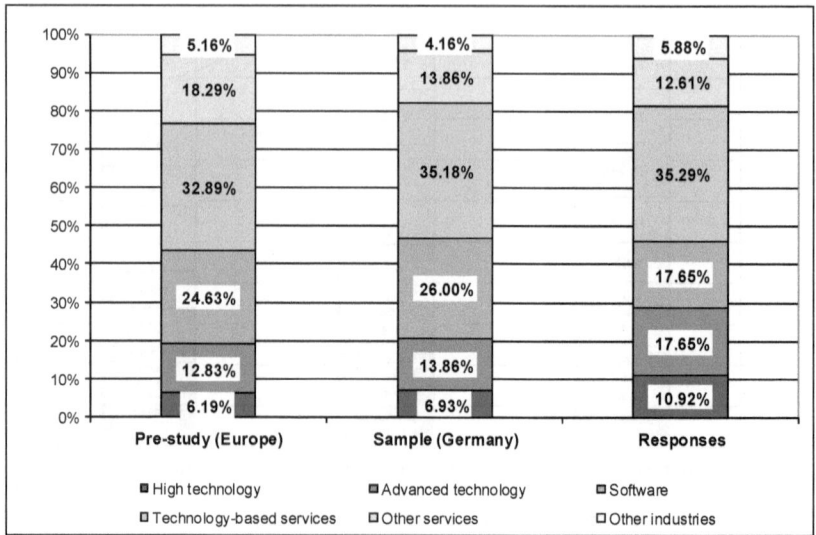

The previous section demonstrated that the characteristics of the group of respondents differ from those of the sample and the population. The differences between the group of respondents and the entire population are critical in terms of representativeness. However, the underlying definition of a research-based spin-off determines the characteristics of the population. In the next step, a non-response-bias test was conducted according to the procedure of Armstrong and Overton (1977). This test is based on the assumption that respondents who completed the survey later share similarities with non-respondents (Armstrong and Overton, 1977; Zinnbauer and Eberl, 2005). Significant differences between the answers of early and late respondents confirm the presence of a non-response bias (Armstrong and Overton, 1977). Therefore, the answers of the earliest 33 per cent and latest 33 per cent of the respondents were compared via a t-test. No significant differences ($p <$ 0.05) could be identified; thus, it is reasonable to conclude that there is no response bias in the data.

6 Operationalisation and data analysis

The following chapter introduces the measurement models that were used in the empirical study. These measurement tools include scales that describe certain characteristics of the management team, especially prior professional experiences. Information regarding these attributes can be obtained through direct questions, whereas the examined aspects of the strategic decision making processes in firms must be collected indirectly because these processes are not observable. Thus, different latent variables were conceptualised to distinguish formal and informal strategy processes and entrepreneurial strategy making. The measures of firm performance encompass the dimensions of growth and profitability. Finally, several firm characteristics and the perceived level of environmental uncertainty are captured. The second section of this chapter outlines important aspects of the data analysis and discusses content validity, convergent validity, and discriminant validity, and several goodness-of-fit indices.

6.1 Measures

The research model that was discussed in chapter 3 follows a multi-level approach and includes variables on the team, firm, and environmental levels. To test the proposed hypotheses and examine the formulated research questions, the study must provide a definition of the underlying measurement models. The research model examines the relationship between the composition and characteristics of the management teams and the strategic orientation of the companies, and the direct and indirect effects on performance are discussed. First, the measures for the variables pertaining to the characteristics of management teams are introduced. Second, the constructs that reflect the strategic orientation of the companies are described. In the third step, the measures of venture performance are discussed. Finally, the variables concerning firm characteristics and perceived environmental uncertainty are explained.

6.1.1   Executive team characteristics

The measures that pertain to the management team included prior professional experience, educational background, team heterogeneity, composition, and demographic attributes and general characteristics.

The previous *entrepreneurial experience* of executive teams was operationalised in two ways. The first approach was based on the work of Dimov and Shepherd (2005) and Patzelt, zu Knyphausen-Aufseß, and Fischer (2008) and defined entrepreneurial experience as the proportion of managers who had possessed previous entrepreneurial experience before joining a spin-off company. The second measure specified entrepreneurial experience as an indicator variable that captured whether at least one member of an executive team had possessed prior entrepreneurial experience before joining the management team of the spin-off. According to Ucbasaran, Westhead et al. (2010), entrepreneurial experience can be further differentiated between the following firm activities: launching a start-up, holding ownership, experiencing a business failure, and managing a successful venture exit. The variables that reflected prior *management experience* and *industry work experience* were defined in the same ways, with the proportion measure and an indicator variable. These two aspects captured previous work experience in management positions and technical positions.

The measurement models that reflected the educational background of management teams were also derived from the work of Dimov and Shepherd (2005) and Patzelt, zu Knyphausen-Aufseß, and Fischer (2008). The *management education* variable was defined as the proportion of executive team members who possess diplomas or bachelor's, master's, or doctoral degrees in business administration, management, or economics. The *technology education* variable captured educational background in technology-related fields, such as engineering, information, natural science and medicine. These variables were conceptualised as suggested by Dimov and Shepherd (2005) and Patzelt, zu Knyphausen-Aufseß, and Fischer (2008). The overall *level of education* was reflected by the proportion of team members with doctoral degrees.

The next set of variables referred to the composition of management teams. For this study, three variables were included to reflect separate aspects of team composition. First, the *external managers* variable captured the percentage of external executives, also known as surrogate entrepreneurs (in contrast to with academic entrepreneurs), in a management team. Second, the *founder* variable represented the percentage of executives who belonged to the original academic founding team of a spin-off. Finally, the *ownership* variable reflected the percentage of managers who were at least partial owners of a company. The heterogeneity of a group can be discussed from different perspectives. In this study, three variables that reflect the *diversity* of the executive teams in terms of prior entrepreneurial experience, educational backgrounds, and composition were calculated. The heterogeneity measures applied the diversity index that was introduced by Blau (1977), which has proven to be reliable and consistent in prior studies (Bantel and Jackson, 1989). The index accounts for the distribution of team members across certain categories. The maximum value of the heterogeneity index corresponds to a situation with equal proportions of groups in each category. On the contrary, diversity is minimised if all team members demonstrate the same characteristics.

Finally, the demographic attributes and general characteristics of the executive teams included team size, average age, and gender distribution. The *team size* variable captured the current number of executives, and the *team age* variable accounted for the average age of all team members. The *gender* aspect considered the proportion of males among the managing directors.

6.1.2   Strategic decision making

The characteristics of the executive teams and companies can be observed objectively, whereas firm-level strategic orientation cannot be measured directly. The strategic management literature was referenced to obtain a definition of the latent variables. Venkatraman and Grant (1986) emphasised the importance of measurement issues in strategic management research. These authors stated that the quality of the underlying measures dictates the quality of the research. Because strategic management research, which has employed single items rather than construct measures, has been criticised (Boyd, Gove, and Hitt, 2005), an increasing number of scales that are based on multiple items have recently been developed. Nevertheless, many studies devote insufficient attention to validity and reliability concerns.

Podsakoff, Shen, and Podsakoff (2006) identified a considerable number of studies in the strategy literature that contain measurement misspecifications. In most of these cases, the measures of the constructs were treated as reflective measures, although the relationship between the constructs and the items implies formative measures. If measures are determinants or causes of a construct with distinguishable facets that are not substitutable and are not highly correlated, then the measures are formative, and the construct is known as a composite variable (Fornell and Bookstein, 1982; MacCallum and Browne, 1993). In contrast, reflective measures can be viewed as reflections of constructs and usually exhibit high correlations with one another. The indicators are interchangeable, and the removal of one or more of them does not change the meaning of the construct itself, which is also known as a latent variable (MacCallum and Browne, 1993).

The first measurement model in this study should reflect the intensity of formal strategic planning activities on the organisational level. Slevin and Covin (1997) distinguished between planned and emergent qualities of strategy and referred to planned and emergent strategy formation modes. Strategies can be planned and/or can emerge over time. Planned strategies comprise rationality in the decision making process as a key characteristic. Chandler, DeTienne et al. (2011) introduced and validated measurement scales concerning the concepts of effectuation and causation (Sarasvathy, 2001). Causation is comparable with planned strategies (Brews and Hunt, 1999), whereas effectuation is consistent with the concept of emergent strategies. For this study, a reflective latent construct that captures the intensity of strategic planning was conceptualised. This construct consists of four items that were adapted from the strategic measurement literature and adjusted to the underlying empirical setting. Two items were derived from the emergent-to-planned strategy scale (Slevin and Covin, 1997), and two items were adapted from the causation construct that was developed by Chandler, DeTienne et al. (2011). Table 6.1 shows the indicators for this construct. On a seven-point Likert scale, respondents indicated the extent to which they agreed or disagreed with the statements that were presented. Marking (1) indicated strong disagreement with the statement, whereas (7) indicated strong agreement, and (4) indicated that a respondent neither agreed nor disagreed. The remaining scale choices represented differing levels of agreement.

Table 6.1: Measurement model for *strategic planning*

Planning items	
PLAN 1	Formal strategic plans serve as the basis for our competitive analysis.
PLAN 2	We have a clear and consistent vision for where we want to end up with our business.
PLAN 3	We organise and implement control processes to make sure that we meet our objectives.
PLAN 4	Our business strategy is carefully planned before significant competitive actions are taken.

Sources: Slevin and Covin (1997), Chandler, DeTienne et al. (2009)

In addition to the level of strategic planning activities, the emphasis of firms on long-term planning and forecasting should be captured. Venkatraman (1989) introduced different dimensions of strategic orientation, including the construct of *futurity*, which also encompasses temporal considerations in the decision making process. In this context, companies pursue actions that are designed to positively influence future firm performance.

This construct has been used and validated in prior research (e.g., Morgan and Strong, 1998, 2003). For this study, the construct was adapted to the empirical setting of the underlying analysis and defined as a latent variable that included four indicators on a seven-point Likert scale. The construct is displayed in table 6.2.

Table 6.2: Measurement model for *futurity*

Futurity items	
FUT 1 (reverse)	Our criteria for resource allocation generally reflect short-term considerations.
FUT 2	Forecasting key indicators of operations is common in our company.
FUT 3	Formal tracking of significant general trends is common in our company.
FUT 4	We often conduct "what if" analyses of critical issues.

Sources: Venkatraman (1989), Morgan and Strong (1998, 2003)

In contrast with the previous two constructs, which more broadly focus on formal planning activities and refer to the rationality aspect of the decision making process, the following variables correspond to an informal and more intuitive, incremental approach to strategising. The theoretical concepts of *flexibility*, *experimentation*, and *partnering* were introduced by Sarasvathy (2001) as distinct dimensions of the overall effectuation concept. Chandler, DeTienne et al. (2011) developed and empirically validated measurement models for these variables. This study specified flexibility, experimentation, and partnering as three latent constructs. The measures were derived from the work of Chandler, DeTienne et al. (2011) and modified to fit the underlying research model. Flexibility was defined by four items and reflects the extent to which a company remains flexible, for example, in terms of opportunity exploitation or resource allocation. The experimentation variable was conceptualised by two items and was designed to measure the level of experimentation in terms of the development of the product and service portfolios of the ventures. The partnering construct captured the ways in which the companies use pre-commitments and collaborations with customers and suppliers to cope with an unpredictable future environment. The models are presented in table 6.3.

Table 6.3: Measurement models for *flexibility*, *experimentation*, and *partnering*

Flexibility items	
FLEX 1	We allow the business to evolve as opportunities emerge.
FLEX 2	We adapt what we are doing to the resources we have.
FLEX 3	We are flexible and take advantage of opportunities as they arise.
FLEX 4	We avoid courses of action that restrict our flexibility and adaptability.
Experimentation items	
EXP 1	We experiment with different products and/or services.
EXP 2	The products/services we now provide are substantially different than we first imagined.
Partnering items	
PART 1	We use pre-commitments from customers and suppliers as often as possible.
PART 2	We use agreements with customers, suppliers, and others to reduce the amount of uncertainty.

Source: Chandler, DeTienne et al. (2011)

The following paragraph introduces the measurement models for the five dimensions of the entrepreneurial orientation concept. The scales that are used are based on previous studies that have developed and tested the measures in terms of validity and reliability (e.g., Khandwalla, 1976; Miller, 1983; Covin and Slevin, 1989; Covin and Covin, 1990; Lumpkin and Dess, 2001; Naldi, Nordquist et al., 2007). The measures for *innovativeness* consisted of three items that were adapted from the work of Covin and Slevin (1989) and Naldi, Nordquist et al. (2007). *Proactiveness* was measured with three items. Two of these items were developed by Covin and Slevin (1989), and the third item was drawn from the study of Lumpkin and Dess (2001). The construct for the *risk-taking* dimension included three items that were adapted from the work of Covin and Slevin (1989) and Naldi, Nordquist et al. (2007). *Competitive aggressiveness* was captured by two indicators: the first indicator was developed by Covin and Slevin (1989), who referred to Khandwalla (1976), and the second indicator was based on the work of Lumpkin and Dess (2001). *Autonomy* was measured by three items that were based on the study of Lumpkin and Dess (1996). The respondents indicated which of two contrasting statements more accurate reflected their companies on a seven-point Likert scale. Marking (1) indicated strong agreement with the first statement, whereas (7) indicated strong agreement with the second statement, and (4) indicated that both statements are equally applicable. The remaining scale choices represented differing levels of agreement with one of the two statements. Generally, higher scores indicated a higher entrepreneurial posture. Table 6.4 provides an overview of the measurement models.

Table 6.4: Measurement models for *entrepreneurial orientation*

Innovativeness items		
INNO 1	Generally, our company prefers to strongly emphasise the marketing of tried and true products or services.	Generally, our company prefers to strongly emphasise R&D, technological leadership, and innovations.
INNO 2	In the last 3 years, our company has introduced no new lines of products or services.	In the last 3 years, our company has introduced very many new lines of products or services.
INNO 3	Changes in products or services have been mostly of a minor nature.	Changes in products or services have usually been quite dramatic.
Proactiveness items		
PRO 1	Typically, we respond to actions which competitors initiate.	Typically, we initiate actions which competitors then respond to.
PRO 2	Our company is very seldom the first one to introduce new products/ services, operating technologies, etc.	Our company is very often the first one to introduce new products/services, operating technologies, etc.
PRO 3	Our company has a strong tendency to "follow the leader" in introducing new products/services or ideas.	Our company has a strong tendency to be ahead of other competitors in introducing novel ideas or products/services.
Risk-taking items		
RISK 1	Generally, our company has a strong tendency towards projects with low risk (with normal and certain results).	Generally, our company has a strong tendency towards high-risk projects (with a chance for high results).
RISK 2	We generally believe, that it is best to achieve our objectives gradually via cautious, incremental behaviour.	We generally believe, that it is best to achieve our objectives via bold and wide-ranging actions.
RISK 3 (reverse)	We typically adopt a fearless, aggressive position in order to maximise the probability of exploiting opportunities.	We typically adopt a "wait-and-see" position in order to minimise the probability of making costly decisions.

Table 6.4: Measurement models for entrepreneurial orientation (continued)

Competitive aggressiveness items		
AGG 1	Normally, we seek to avoid competitive clashes, preferring a "live-and-let-live" posture.	Normally, we adopt a very competitive, "beat-the-competitor" posture.
AGG 2 (reverse)	Our actions towards other competitors can be termed aggressive.	Our actions towards other competitors can be termed accommodating.
Autonomy items		
AUT 1 (reverse)	In our company, very many changes suggested by our employees are implemented.	In our company, very few changes suggested by our employees are implemented.
AUT 2	Identifying new products and services is the responsibility of a small number of individuals.	Identifying new products and services is done by all members of the company.
AUT 3	Our company discourages independent activity to develop new services and products.	Our company encourages independent activity to develop new services and products.

Sources: Covin and Slevin (1989), Lumpkin and Dess (2001), Naldi, Nordquist et al. (2007)

6.1.3 Firm performance

Performance can be measured in different ways. Generally, subjective and objective measures can be differentiated (Dess and Robinson, 1984; Sapienza, Smith and Gannon, 1988). Subjective measures are based on the perceptions of respondents regarding the performance of their companies. Objective measures are based on financial figures, such as growth in turnover, employees, or profit. Using objective measures in the context of high technology small firms might be problematic because these data are usually not accessible and often do not reflect firm performance appropriately. For example, growth rates can be high in an early life-cycle stage if a company begins as a small entity. Nevertheless, subjective measures can have biases and measurement errors. However, prior research noted that subjective performance indicators are capable of reliably reflecting objective measures. Chandler and Hanks (1993) proved that estimations with regard to performance aspects, such as earnings or growth, are highly correlated with archival data. The use of subjective measures is also expected to yield enhanced response rates. Generally, previous studies have distinguished and contrasted growth and profitability (Kirchhoff, 1977). Along these lines, Wiklund and

Shepherd (2003) stated that performance is a multidimensional phenomenon that can be captured by comparing the performance of a company to that of its competitors. Therefore, in this study, the performance dimensions of growth and profitability were considered and based on self-perceived estimations. The respondents were asked to compare the performance of their companies to the performance of their competitors. The seven-point Likert scale on which they responded ranged from "much worse" to "much better". Referring to the works of Chandler and Hanks (1993) and Wiklund and Shepherd (2003), this study measured growth by three items and profitability by two indicators.

Table 6.5: Measurement models for firm performance

Performance items	
GROWTH 1	Growth in sales
GROWTH 2	Growth in employees
GROWTH 3	Growth in market share
PROFIT 1	Net profits
PROFIT 2	Return on investment

Sources: Chandler and Hanks (1993), Wiklund and Shepherd (2003)

6.1.4   Firm characteristics and environmental uncertainty

The maturity of the companies was measured by three discriminative variables. The firm age was calculated by the number of years that had passed since the firms were formally incorporated. The firm size was measured by the number of employees within each company. Additionally, following the work of Kazanjian (1988), Galbraith and Vesper (1982), Chandler and Hanks (1993), and Hanks, Watson et al. (1993), this study distinguished five life-cycle stages: start-up, expansion, maturity, diversification, and decline. On the basis of a brief description of each phase, the respondents selected the development stage that most apply corresponded to the current situation of their companies. The different stages are displayed in table 6.6.

Table 6.6: Conceptualisation of life-cycle stages

Life-cycle stage	Characteristics
Start-up	Formal establishment, informal structure, focus on obtaining resources and building prototypes, inconsistent growth rates
Expansion	Functionally organised structure emerges, early formalisation of policies, focus on volume production, capacity expansion, rapid growth rate
Maturity	More bureaucratic organisation, focus on making business profitable, expense control, establishing management system, slow growth rate
Diversification	Divisionalisation, diversification of products and market scope, use of sophisticated controls and planning system, high growth rate
Decline	Demand for products levels off, low rate of innovation, profitability starts to drop, redefinition of mission and strategy, declining growth rate

Sources: Kazanjian (1988), Galbraith and Vesper (1982), Chandler and Hanks (1993), and Hanks et al. (1993)

The spin-off companies in the sample differed with regard to their portfolios and their underlying research and technology bases, and such differences could have influenced the dependent variable. Therefore, three variables were included in the analysis to measure whether the current portfolio of a company predominantly consists of products or services and the extent to which the portfolio is based on extensive research efforts and high technology. The discussion of rationality and planning in strategic decision making processes is linked to the uncertain and unpredictable conditions of the environment. Environmental uncertainty can be measured objectively or subjectively (Boyd, Dess, and Rasheed, 1993). In this study, self-perceived measures were used because the decisions of executive teams are primarily influenced by their perceptions of the environment. The study utilised measures for state and effect uncertainty (Milliken, 1987) that were adapted from the work of McKelvie, Haynie, and Gustavsson (2011) and that reflected the level of environmental uncertainty. State uncertainty refers to certain factors of the environment, such as demographic shifts, that cannot be predicted, whereas effect uncertainty reflects the inability to estimate the ways in which changes in the environment can affect a company (McKelvie, Haynie, and Gustavsson, 2011). Each variable was measured by two composite items. On a seven-point Likert scale, respondents chose which of the presented statements was more applicable to their companies. Table 6.7 shows the different scenarios.

Table 6.7: Conceptualisation of environmental uncertainty

State uncertainty items		
UNC 1	The demand for our products/services is steady over time.	The rate of demand for our products/services fluctuates significantly.
UNC 2 (reverse)	Future technological innovations affecting the viability of our products/services are likely to be frequent and major.	Future technological innovations affecting the viability of our products/services are likely to be rare and minor.
Effect uncertainty items		
UNC 3	We can predict our customers' preferences and demands with regard to our products/services over time.	It is not possible to predict in advance demand changes regarding our products/services.
UNC 4 (reverse)	It is not possible to predict kinds or timing of future innovations that will affect the viability of our products/services.	We can predict the nature and source of innovations that affect the viability of our products/services.

Source: McKelvie, Haynie, and Gustavsson (2009)

6.2 Data analysis

The postulated hypotheses and formulated research questions were examined through the use of bivariate and multivariate analyses. Because the strategic orientation dimensions could not be measured directly and were operationalised as latent variables, structural equation models were developed. The *Statistical Package for Social Science SPSS 18* and the software programme *Analysis of Moment Structures AMOS 18* were used to conduct the analyses. Prior to the discussion of the results, the fit of each model must be evaluated both on the local level, regarding the applied constructs, and on the global level, concerning the overall fit of the models. First, the fit of the measurement models should be discussed. Both newly designed and established measurement scales should be evaluated prior to use (Bortz and Döring, 2006). In this context, different criteria for estimating reliability and validity must be considered. Reliability reflects the internal consistency of the latent variables in the model, whereas construct validity captures whether the variables actually measure the intended construct. In this regard, the following aspects must be mentioned: (1) content validity is usually estimated by experts in the field who are competent in evaluating the plausibility of such constructs; (2) convergent validity requires a high correspondence between the scores of two or more different measures of the same construct; (3) discriminant validity assumes that the operationalisation of a construct is actually a discrete model that is not related to other theoretically possible models.

All of the measurement models were drawn from previous research, and the required modifications for the underlying empirical setting were discussed with experts. Thus, content validity is assumed for the latent constructs in this study. To test for reliability, convergent validity, and discriminant validity, this study utilised the following criteria. Internal consistency was estimated by calculating Cronbach's alpha and item-to-total correlations. The explorative and confirmatory factor analysis assessed the explained variance and factor loadings as well as the factor reliability, average variance extracted (AVE), and indicator reliabilities. Table 6.8 provides an overview of the expected ranges for the factors.

Table 6.8: Fit indices for measurement models

Analysis	Fit index	Requirement
Internal consistency	Cronbach's alpha	$\geq 0.7^1$
	Item-to-total correlation	≥ 0.5
Explorative factor analysis	Explained variance	≥ 0.5
	Factor loading	≥ 0.5
Confirmatory factor analysis	Factor reliability	≥ 0.6
	AVE	≥ 0.5
	Indicator reliability	≥ 0.4
Discriminant validity	χ^2-difference test	≥ 3.841
	Fornell-Larcker criterion	$AVE(\xi_j) > r^2(\xi_i, \xi_j)$

[1] if fewer than 4 indicators: 0.4 (Peter, 1997)

Sources: Nunnally (1978), Fornell and Larcker (1981), Bagozzi and Yi (1988), Bearden, Netemeyer, and Teel (1989), Bagozzi and Baumgartner (1994), Peter (1997), Homburg (2000), Backhaus, Erichson et al. (2003), Homburg, Pflesser, and Klarmann (2008)

Cronbach's alpha is often used to capture the reliability or internal consistency of a measurement model. As the covariances or correlations between the indicators increase, this value becomes closer to 1. The range that reflects an acceptable level of reliability is controversial. Peter (1997) considers 0.4 to be a satisfactory score for constructs with fewer than four indicators, whereas Nunnally (1978) states that Cronbach's alpha should be at least

0.7. The item-to-total correlation represents the correlation of an indicator variable with the sum of all indicators that belong to the same factor. If the item-to-total correlation of an indicator is not sufficient, then the item can be excluded from the construct. Each indicator of the latent variable should have a minimum item-to-total correlation of 0.5 (Bearden, Netemeyer, and Teel (1989). Subsequently, an explorative factor analysis was conducted with the reduced scales to determine whether the indicators form a single factor, which indicates convergent validity. The factor should explain at least 50 per cent of the variance of the corresponding indicators (Peter, 1997), and factor loadings should not be less than 0.5 for each indicator (Backhaus, Erichson et al. (2003).

In the next step, a confirmatory factor analysis was conducted; this analysis postulates a structure with one factor and tests for reliability. The reliability of an indicator generally describes whether it properly measures the underlying latent construct. Indicator reliability is defined as the percentage of variance of an indicator that is explained by the factor to which it belongs. Thus, the value is between 0 and 1 and should be at least 0.4 (Bagozzi and Baumgartner, 1994). In large samples (n > 1,000), the minimum value should lie between 0.1 and 0.2 (Balderjahn, 1986). Both the factor reliability and the AVE demonstrate the adequacy of the measurement of the latent construct by the corresponding indicators. Factor reliability reports the total variance that is captured by the model and should not be below 0.6, whereas the limit for the AVE is 0.5 (Bagozzi and Yi, 1988). Discriminant validity can be assessed by applying the χ^2 difference test, which compares the original model with a new model, in which the correlation between two latent variables is fixed to 1. The χ^2 difference test is intended to yield a score of at least 3.841 (Homburg, 2000). The discriminant validity of the underlying constructs is also evaluated by the Fornell/Larcker criterion, which is generally stricter and requires that the AVE of a factor is higher than the squared correlations of this factor with all other factors in the model (Fornell and Larcker, 1981). In the second step, different indices with regard to the global model are discussed. The recommended thresholds are displayed in table 6.9.

Inferential statistical criteria, such as the root mean squared error of approximation (RMSEA), estimate the goodness of fit based on statistical tests. The RMSEA indicates the closeness of a model to the observed reality and should be not higher than 0.1 (Browne and Cudeck, 1993). The χ^2 value can be used in combination with the number of degrees of freedom as an appropriate fit index. A good model fit is indicated if the relation of the χ^2 value and the number of degrees of freedom does not exceed 3.0 (Homburg and Giering, 1996). Incremental goodness-of-fit indices examine how the model fit improves following the transition from the base model to the relevant model. In the base model, which does not offer any plausibility with regard to content, all of the variables are assumed to be uncorrelated with one another. The normed fit index (NFI) relates the χ^2 values of the relevant model to the

χ^2 values of the base model and ranges from 0 to 1, where a score of 0.9 or higher suggests an acceptable model fit (Bentler and Bonnett, 1980). The comparative fit index (CFI) also considers the number of degrees of freedom and should also have a minimum value of 0.9 (Homburg and Baumgartner, 1995).

Table 6.9: Goodness-of-fit indices

Fit index	Requirement
χ^2/df	≤ 3.0
P	≥ 0.1
RMSEA	≤ 0.1
NFI	≥ 0.9
CFI	≥ 0,9
PCLOSE	≥ 0,1

Sources: Bentler and Bonnett (1980), Browne and Cudeck (1993), Homburg and Giering (1996), Homburg and Baumgartner (1995)

7 Empirical results

In this chapter, the results for the examined links between team characteristics and performance, between strategic decision making and performance, and between team characteristics and strategic decision making are presented. Both formal and informal strategy approaches and the level of entrepreneurial posture are analysed. In addition, the study examines several firm and environmental attributes and their impacts on strategic decision making. The discussion first considers each relationship separately to demonstrate direct effects. This analysis includes a critical evaluation of the measurement models based on appropriate goodness-of-fit indices. Second, some factors will be analysed in combination to identify indirect relationships and estimate these factors' relative importance.

7.1 Link between executive characteristics and firm performance

This section describes the relationships between executive team characteristics and firm performance. First of all, the two performance dimensions that are considered in this study – growth and profitability – are tested in terms of the fit of the measurement models. Because content validity was already discussed in the previous chapter, the following sections focus on convergent and discriminant validity. As outlined in chapter 4, the following explanatory variables are examined: prior professional experience, educational background, team composition and heterogeneity, and demographic attributes and general characteristics. Each aspect includes several variables. Table 7.1 provides an overview of the variables used in this study (means, standard deviations) and a correlation matrix. Bivariate and multivariate analyses examine both the direct impact of each factor and its impact combined with other factors. This section reports significant results, effect sizes, and explained variance.

Table 7.1: Descriptive statistics and correlations

Variables	Means	SD	1	2	3	4	5	6	7	8	9	10	11	12	13	14
1 Entrepreneurial experience	0.46	0.499														
2 Management experience	0.51	0.501	0.372**													
3 Industry work experience	0.49	0.501	0.118	0.296**												
4 Management education	0.17	0.288	0.182*	0.079	0.049											
5 Technology education	0.80	0.326	-0.135	-0.127	0.047	-0.498**										
6 Level of education	0.64	0.382	-0.048	-0.074	-0.054	-0.245**	0.278**									
7 Diversity experience	0.15	0.156	0.555**	0.593**	0.388**	0.202**	-0.222**	-0.131								
8 Diversity education	0.12	0.183	0.327**	0.318**	0.087	0.503**	-0.411**	-0.265**	0.431**							
9 Diversity composition	0.11	0.154	0.289**	0.306**	0.189*	0.065	-0.126	-0.101	0.441**	0.339**						
10 External manager	0.18	0.316	0.274**	0.313**	0.249**	0.180**	-0.125	-0.042	0.389**	0.254**	0.638**					
11 Ownership	0.80	0.349	-0.230**	-0.263**	-0.059	-0.267**	0.113	0.146*	-0.217**	-0.219**	-0.375**	-0.325**				
12 Founder	0.76	0.327	-0.215**	-0.286**	-0.098	-0.207**	0.210**	0.147*	-0.314**	-0.419**	-0.563**	-0.591**	0.321**			
13 Team size	2.30	1.016	0.257**	0.356**	0.143*	0.062	-0.098	-0.135	0.440**	0.394**	0.540**	0.428**	-0.224**	-0.415**		
14 Average age	42.7	11.7	0.093	0.352**	0.186*	0.025	-0.225**	0.193**	0.143	0.057	0.120	0.149	-0.098	-0.126	0.039	
15 Gender distribution	0.92	0.216	0.040	0.117	0.022	-0.088	0.084	0.114	0.005	-0.088	-0.029	-0.025	0.082	0.109	-0.024	0.077

N = 193

*p < 0.05; **p < 0.01; ***p < 0.001

7.1.1   Reliability and validity of performance measures

In the following sections, the fit of the two measurement models for growth and profitability is discussed. Table 7.2 shows the indicators of convergent validity of the measurement model for the performance dimension of growth (e.g., internal consistency and explorative and confirmatory factor analysis). The growth variable is conceptualised by three reflective items. Overall, the indicators demonstrate a very good fit of the measurement model. The Cronbach's alpha value of 0.876 shows good internal consistency, and all indicators have an item-to-total correlation of at least 0.720. According to explorative factor analysis, 80.1 per cent of the variance of the indicators is explained, and all factor loadings are above 0.872. Confirmatory factor analysis yields relatively high factor reliability of 0.881 and AVE of 0.713. In terms of indicator reliability, a minimum of 59.9 per cent of the variance of each indicator is explained by the underlying factor growth.

Table 7.2: Fit indices for latent variable growth

Analysis	Fit index	Result
Internal consistency	Cronbach´s alpha (≥ 0.7)	0.876
	Item-to-total (≥ 0.5) GROWTH 1 GROWTH 2 GROWTH 3	0.822 0.720 0.743
Explorative factor analysis	Explained variance (≥ 0.5)	0.801
	Factor loading (≥ 0.5) GROWTH 1 GROWTH 2 GROWTH 3	0.927 0.872 0.885
Confirmatory factor analysis	Factor reliability (≥ 0.6)	0.881
	AVE (≥ 0.5)	0.713
	Indicator reliability (≥ 0.4) GROWTH 1 GROWTH 2 GROWTH 3	0.882 0.599 0.649

N = 177

The model measuring profitability also demonstrates a good fit in terms of convergent validity. The latent variable consists of two items. The Cronbach's alpha value of 0.807 indicates high covariance between the indicators, which all have sufficient item-to-total correlations. Explorative factor analysis reveals that the explained variance accounts for 83.9

per cent of the total variance of the indicators. Also, the high factor loadings confirm a good model fit. Confirmatory factor analysis yields factor reliability of 0.826 and AVE of 0.707, which are both notably above the required thresholds. The profitability factor explains 51.3 per cent and 92.2 per cent, respectively, of the variance of the two indicators, which is satisfactory. An overview of the results is displayed in table 7.3.

Table 7.3: Fit indices for latent variable profitability

Analysis	Fit index	Result
Internal consistency	Cronbach's alpha (≥ 0.7)	0.807
	Item-to-total (≥ 0.5) PROFIT 1 PROFIT 2	 0.678 0.678
Explorative factor analysis	Explained variance (≥ 0.5)	0.839
	Factor loading (≥ 0.5) PROFIT 1 PROFIT 2	 0.916 0.916
Confirmatory factor analysis	Factor reliability (≥ 0.6)	0.826
	AVE (≥ 0.5)	0.707
	Indicator reliability (≥ 0.4) PROFIT 1 PROFIT 2	 0.922 0.515

N = 172

The following discriminant analysis determines whether the two constructs reflect distinct performance dimensions. The discriminant validity of the underlying constructs was evaluated by the Fornell-Larcker criterion, which requires the AVE of a factor to be higher than the squared correlations of this factor with all other factors in the model. Table 7.4 reports satisfactory results.

Table 7.4: Fornell-Larcker criterion for growth and profitability

	GROWTH	PROFIT
GROWTH	0.713	-
PROFIT	0.320	0.707

Furthermore, discriminant validity is tested on the indicator level. Table 7.5 shows the correlations between the indicators and the constructs. All indicators demonstrate the highest correlation with the construct to which they belong according to the model specification.

Table 7.5: Discriminant validity on the indicator level for growth and profitability

	GROWTH	PROFIT
GROWTH 1	**0.925**	0.524
GROWTH 2	**0.781**	0.442
GROWTH 3	**0.817**	0.462
PROFIT 1	0.544	**0.960**
PROFIT 2	0.405	**0.716**

In summary, the goodness-of-fit indices demonstrate satisfactory results for both performance dimensions growth and profitability. The tests for discriminant validity indicate that the two variables reflect different aspects of performance.

7.1.2 Prior professional experience and firm performance

As outlined in chapter 6, previous professional experience is assessed on the team level following two different approaches. The first measure captures whether at least one member of the executive team had specific experience before joining the spin-off. The second one indicates the proportion of managers who have these types of experience. Table 7.6 presents a descriptive overview of the variables and reveals some noteworthy insights. In approximately half of the companies, at least one member of the management team has prior entrepreneurial, management, or industry work experience. On average, only approximately one third of the team members have prior experience in one of the studied areas. These figures already indicate a lack of prior professional experience in the majority of executive teams.

The Kolmogorov-Smirnov test indicates that the variables are not exactly normally distributed. In larger samples, the gap between the observed distribution and an assumed normal distribution, which almost always exists to some degree, is usually identified. Thus, the value of significance tests in this context is questionable. Arbuckle (2007) mentioned that it is more useful to identify serious discrepancies from a normal distribution. In this respect, Kline (2005) recommended that the absolute values of skewness and kurtosis should not be higher than 3 and 10, respectively.

Table 7.6: Descriptive statistics of prior professional experience

	N	Mean	SD	KS	P (KS)	Skewness	Kurtosis
Entrepreneurial experience 1	193	0.46	0.499	5.049	0.000	0.178	-1.989
Management experience 1	193	0.51	0.501	4.826	0.000	-0.052	-2.018
Industry work experience 1	193	0.49	0.501	4.789	0.000	0.031	-2.020
Entrepreneurial experience 2	193	0.28	0.362	4.507	0.000	0.976	-0.432
Management experience 2	193	0.33	0.380	4.165	0.000	0.685	-0.996
Industry work experience 2	193	0.33	0.388	4.283	0.000	0.732	-0.982

SD: Standard deviation, KS: Kolmogorov-Smirnov-test

Tables 7.7 and 7.8 show the results regarding the impact of prior professional experience on firm performance. In models 1 to 3 of each table, the effect size is measured separately for each variable, whereas model 4 contains the full model including all of the variables. Hypotheses 1.1.1-1.1.3 predicted a positive impact of prior entrepreneurial, management, and industry work experience on firm performance. In the first approach, only management experience has a statistically significant positive impact on growth (β = 0.178, p < 0.05), demonstrating a small effect size and explaining 3.2 per cent of the variance of the dependent variable. Prior entrepreneurial and industry work experience have only marginally positive impacts on growth, with small effect sizes for both entrepreneurial experience (β = 0.139, p < 0.1) and industry work experience (β = 0.142, p < 0.1). These factors explain 1.9 and 2.0 per cent of the variance, respectively. In the second approach, which captures the percentage of managers who have had specific experiences, no significant results could be identified. In the full model, none of the variables remain significant, which is presumably due to correlations between the explanatory variables. The fit of the models is good.

Table 7.7: Prior professional experience 1 and growth

	Model 1	Model 2	Model 3	Model 4
Entrepreneurial experience 1	$0.139^{\dagger}$			0.081
Management experience 1		0.178*		0.120
Industry work experience 1			$0.142^{\dagger}$	0.098
R^2	0.019	0.032	0.020	0.047
				1.481
χ^2	0.6	3.7	3.6	8.9
df	2	2	2	6
P (≥ 0.1)	0.733	0.160	0.164	0.180
χ^2/df (≤ 3.0)	0.311	1.835	1.805	1.481
RMSEA (≤ 0.1)	0.000	0.066	0.065	0.050
CFI (≥ 0.9)	1.000	0.944	0.994	0.991
NFI (≥ 0.9)	0.998	0.988	0.988	0.974
PCLOSE (≥ 0.1)	0.823	0.296	0.302	0.431

†p < 0.1; *p < 0.05; **p < 0.01; ***p < 0.001

Table 7.8: Prior professional experience 2 and growth

	Model 1	Model 2	Model 3	Model 4
Entrepreneurial experience 2	0.117			0.108
Management experience 2		0.036		0.029
Industry work experience 2			0.105	0.100
R^2	0.014	0.001	0.011	0.024
χ^2	0.7	2.0	3.6	6.8
df	2	2	2	6
P (≥ 0.1)	0.706	0.373	0.165	0.340
χ^2/df (≤ 3.0)	0.348	0.985	1.805	1.133
RMSEA (≤ 0.1)	0.000	0.000	0.065	0.026
CFI (≥ 0.9)	1.000	1.000	0.994	0.997
NFI (≥ 0.9)	0.998	0.993	0.988	0.977
PCLOSE (≥ 0.1)	0.804	0.530	0.302	0.612

†p < 0.1; *p < 0.05; **p < 0.01; ***p < 0.001

In the next step, the relationship between prior professional experience and profitability of the company is analysed. Using the first approach, measuring team experience with an indicator variable, only prior industry work experience was found to have a statistically significant impact ($\beta = 0.191$, $p < 0.05$), whereas entrepreneurial and management experience each had no effect. The reported effect size is small, and the factor explains 3.7 per cent of the variance of the dependent variable. The impact remains significant in the full model. The fit indices can only be estimated for the full model because the models measuring the single impacts are fully identified. In the full model, all the criteria are fulfilled. In the second approach, which measures the proportion of team members having prior experience, none of the variables was found to be statistically significantly related to the dependent variable. Entrepreneurial experience has a marginally significant impact on profitability ($\beta = 0.129$, $p < 0.1$), with a small effect size, explaining 1.7 per cent of the total variance. This effect remains marginally significant in the full model, which has adequate fit indices.

Table 7.9: Prior professional experience 1 and profitability

	Model 1	Model 2	Model 3	Model 4
Entrepreneurial experience 1	0.036			0.044
Management experience 1		0.040		-0.007
Industry work experience 1			0.191*	0.184*
R^2	0.001	0.002	0.037	0.037
χ^2				0.6
df	0	0	0	2
P (≥ 0.1)				0.725
χ^2/df (≤ 3.0)				0.321
RMSEA (≤ 0.1)				0.000
CFI (≥ 0.9)				1.000
NFI (≥ 0.9)				0.996
PCLOSE (≥ 0.1)				0.817

[†]p < 0.1; *p < 0.05; **p < 0.01; ***p < 0.001

Table 7.10: Prior professional experience 2 and profitability

	Model 1	Model 2	Model 3	Model 4
Entrepreneurial experience 2	$0.129^{\dagger}$			$0.145^{\dagger}$
Management experience 2		0.024		-0.041
Industry work experience 2			0.137	0.135
R^2	0.017	0.001	0.019	0.036
χ^2				0.7
df	0	0	0	2
P (≥ 0.1)				0.345
χ^2/df (≤ 3.0)				0.708
RMSEA (≤ 0.1)				0.000
CFI (≥ 0.9)				1.000
NFI (≥ 0.9)				0.995
PCLOSE (≥ 0.1)				0.805

$^{\dagger}p < 0.1$; $^{*}p < 0.05$; $^{**}p < 0.01$; $^{***}p < 0.001$

7.1.3 Educational background and firm performance

The impact of educational background was examined from different perspectives. The management education variable encompasses executives who hold a university or equivalent degree in fields such as business administration, management, and economics. Degrees in technology-related fields, such as engineering, informatics, or natural science, are covered by the technology education variable. The overall level of education is defined as the proportion of managers who have a doctoral degree. Descriptive statistics concerning educational background are displayed in table 7.11. On average, only 17.3 per cent of the executive team members have a degree in management-related fields, whereas 79.5 per cent have a degree in a technical area. The educational level in the sample is generally very high, as indicated by the average proportion of 63.6 per cent of team members who hold a doctoral degree. The significance test does not confirm exact normal distributions of the variables, but, as outlined above, the absolute values for skewness and kurtosis are in an acceptable range and do not indicate serious violations of the normal distribution assumption.

Table 7.11: Descriptive statistics of educational background

	N	Mean	SD	KS	P (KS)	Skewness	Kurtosis
Management education	193	0.173	0.288	5.475	0.000	1.644	1.818
Technology education	193	0.795	0.326	5.317	0.000	-1.447	0.849
Level of education	193	0.636	0.382	3.825	0.000	-0.518	-1.173

SD: Standard deviation, KS: Kolmogorov-Smirnov-test

The results regarding the relationships between the educational background of the executive team and firm performance are displayed in tables 7.12 and 7.13. Models 1 and 2 test for the impact of education on management or technology, whereas model 3 uses the percentage of team members who hold a doctoral degree. In chapter 4, no concrete hypotheses regarding the effect size were postulated. Only in the full model is a positive relationship between technology degree and growth shown. The goodness-of-fit indices are in a satisfactory range.

Table 7.12: Educational background and growth

	Model 1	Model 2	Model 3	Model 4
Management education	0.036			0.085
Technology education		0.083		$0.150^{\dagger}$
Level of education			-0.068	-0.090
R^2	0.001	0.007	0.005	0.021
χ^2	2.0	3.8	2.2	7.7
df	2	2	2	6
P (≥ 0.1)	0.373	0.886	0.341	0.259
χ^2/df (≤ 3.0)	0.985	0.152	1.077	1.286
RMSEA (≤ 0.1)	0.000	0.068	0.020	0.039
CFI (≥ 0.9)	0.993	0.994	0.999	0.995
NFI (≥ 0.9)	1.000	0.987	0.993	0.979
PCLOSE (≥ 0.1)	0.530	0.286	0.498	0.529

$^{\dagger}p < 0.1$; $^{*}p < 0.05$; $^{**}p < 0.01$; $^{***}p < 0.001$

Table 7.13: Educational background and profitability

	Model 1	Model 2	Model 3	Model 4
Management education	0.037			0.046
Technology education		0.055		0.112
Level of education			-0.027	-0.020
R^2	0.001	0.003	0.001	0.009
χ^2				0.5
df	0	0	0	2
P (≥ 0.1)				0.793
χ^2/df (≤ 3.0)				0.232
RMSEA (≤ 0.1)				0.000
CFI (≥ 0.9)				1.000
NFI (≥ 0.9)				0.997
PCLOSE (≥ 0.1)				0.865

$^{\dagger}p < 0.1$; $^{*}p < 0.05$; $^{**}p < 0.01$; $^{***}p < 0.001$

7.1.4   Heterogeneity and firm performance

Three aspects of team heterogeneity were introduced in the previous chapter: prior professional experience, educational background, and management team composition. Table 7.14 presents a descriptive overview of these aspects. The average level of diversity is highest in the category of previous professional experience, whereas the highest variance is reported for the educational background of the team members. Although the Kolmogorov-Smirnov test does not indicate perfect normal distributions of the variables, the maximum absolute values of 1.118 for skewness and 0.836 for kurtosis do not indicate extreme differences between the observed distributions and the assumed normal distribution. According to table 7.1, all of the heterogeneity measures are significantly positively correlated with each other.

Table 7.14: Descriptive statistics of heterogeneity

	N	Mean	SD	KS	P (KS)	Skewness	Kurtosis
Diversity prior experience	193	0.146	0.156	3.827	0.000	0.637	-0.836
Diversity education	193	0.118	0.183	5.821	0.000	1.118	-0.418
Diversity composition	193	0.110	0.154	5.186	0.000	1.056	-0.238

SD: Standard deviation, KS: Kolmogorov-Smirnov-test

Tables 7.15 and 7.16 report the results concerning the relationship between team heterogeneity and firm performance. Models 1 to 3 take each single aspect into consideration, whereas model 4 includes the full model with all diversity variables. Hypotheses 1.2.1-1.2.3 predict positive effects on firm performance of heterogeneity of prior professional experience, educational background, and the composition of the executive team. The results indicate only a statistically significant relationship between diversity in terms of team composition and firm growth, with a medium effect size ($\beta = 0.224$, $p < 0.01$), thus explaining 5.0 per cent of the variance. The effect remains statistically significant in the full model. However, no statistically significant links are revealed between the other explanatory variables and the dependent variables. Furthermore, all models using profitability as the outcome are not significant. All fit indices fulfil the discussed requirements.

Table 7.15: Heterogeneity and growth

	Model 1	Model 2	Model 3	Model 4
Diversity prior experience	0.113			0.017
Diversity education		0.073		-0.014
Diversity composition			0.224**	0.221*
R^2	0.013	0.005	0.050	0.051
χ^2	0.9	3.4	1.1	5.6
df	2	2	2	6
P ($\geq$ 0.1)	0.650	0.185	0.590	0.471
χ^2/df ($\leq$ 3.0)	0.430	1.688	0.528	0.931
RMSEA ($\leq$ 0.1)	0.000	0.060	0.000	0.000
CFI ($\geq$ 0.9)	1.000	0.995	1.000	1.000
NFI ($\geq$ 0.9)	0.997	0.988	0.996	0.986
PCLOSE ($\geq$ 0.1)	0.762	0.328	0.715	0.724

[†]$p < 0.1$; *$p < 0.05$; **$p < 0.01$; ***$p < 0.001$

Table 7.16: Heterogeneity and profitability

	Model 1	Model 2	Model 3	Model 4
Diversity prior experience	0.020			-0.061
Diversity education		-0.036		-0.051
Diversity composition			0.103	0.049
R^2	0.000	0.001	0.011	0.007
χ^2				2.5
df	0	0	0	2
P ($\geq$ 0.1)				0.281
χ^2/df ($\leq$ 3.0)				1.270
RMSEA ($\leq$ 0.1)				0.038
CFI ($\geq$ 0.9)				0.996
NFI ($\geq$ 0.9)				0.985
PCLOSE ($\geq$ 0.1)				0.437

[†]$p < 0.1$; *$p < 0.05$; **$p < 0.01$; ***$p < 0.001$

7.1.5   Composition and firm performance

Several descriptive statistics regarding the composition of the management teams are displayed in table 7.17. The team composition variables answer the questions of whether the team members joined the team as external managers from outside the research environment, whether they belonged to the original founding team and whether they are owner or co-owner of the company. The distributions are not exactly normally distributed according to the Kolmogorov-Smirnov test. The absolute values of skewness and kurtosis do not exceed 1.530 and 1.020, respectively. Thus, the normal distribution assumption is only moderately violated.

Table 7.17: Descriptive statistics of composition

	N	Mean	SD	KS	P (KS)	Skewness	Kurtosis
External manager	193	0.181	0.316	5.935	0.000	1.530	1.020
Ownership	193	0.803	0.349	6.028	0.000	-1.513	0.752
Founder	193	0.755	0.327	5.052	0.000	-0.974	-0.316

SD: Standard deviation, KS: Kolmogorov-Smirnov-test

The descriptive overview displayed in table 7.17 reveals that, on average, only 18.1 per cent of the executives in a team are external managers who joined the spin-off from outside the research organisation. Furthermore, 80.3 per cent of the team members own a stake in the company, and 75.5 per cent belonged to the original founding team. According to table 7.1, the proportion of external managers is statistically significantly and negatively correlated with the percentage of management team members who are owners and with the percentage who were members of the original founding team.

Hypothesis 1.3 postulated that including external managers on the management team has a positive impact on firm performance. The links between the proportions of owners and founders in the team and firm performance were addressed via research questions. The results displayed in tables 7.18 and 7.19 show no significant relationships. Only in the full model is the percentage of executives who belonged to the original founding team marginally negatively related to growth. The fit indices are all in an acceptable range.

Table 7.18: Composition and growth

	Model 1	Model 2	Model 3	Model 4
External manager	0.000			-0.019
Ownership		0.011		-0.111
Founder			-0.097	-0.191[†]
R^2	0.000	0.000	0.009	0.017
χ^2	0.6	0.9	1.1	2.4
df	2	2	2	6
P ($\geq$ 0.1)	0.729	0.653	0.585	0.885
χ^2/df ($\leq$ 3.0)	0.317	0.426	0.536	0.392
RMSEA ($\leq$ 0.1)	0.000	0.000	0.000	0.000
CFI ($\geq$ 0.9)	1.000	1.000	1.000	1.000
NFI ($\geq$ 0.9)	0.998	0.997	0.996	0.996
PCLOSE ($\geq$ 0.1)	0.820	0.764	0.712	0.960

[†]$p < 0.1$; *$p < 0.05$; **$p < 0.01$; ***$p < 0.001$

Table 7.19: Composition and profitability

	Model 1	Model 2	Model 3	Model 4
External manager	-0.059			-0.014
Ownership		-0.042		-0.068
Founder			-0.021	-0.105
R^2	0.004	0.002	0.000	0.005
χ^2				0.6
df	0	0	0	2
P ($\geq$ 0.1)				0.731
χ^2/df ($\leq$ 3.0)				0.314
RMSEA ($\leq$ 0.1)				0.000
CFI ($\geq$ 0.9)				1.000
NFI ($\geq$ 0.9)				0.999
PCLOSE ($\geq$ 0.1)				0.821

[†]$p < 0.1$; *$p < 0.05$; **$p < 0.01$; ***$p < 0.001$

7.1.6   General attributes and firm performance

Table 7.20 gives a statistical overview of several demographic attributes and general characteristics of the management team. The mean executive team size is larger in the group of respondents (2.3) than in the original population (1.7). The average age of all managers in the examined spin-off companies is 42.7 years. As shown by the gender distribution, the executive teams are dominated by males and include, on average, only 9.2 per cent females. The team size and average age variables demonstrate severe violations of the normal distribution assumption. Thus, the results have to be discussed with reservation.

Table 7.20: Descriptive statistics of general attributes

	N	Mean	SD	KS	P (KS)	Skewness	Kurtosis
Team size	193	2.30	1.016	4.360	0.000	1.220	3.173
Average age	120	42.7	11.692	1.340	0.055	1.976	12.379
Gender distribution	193	0.92	0.216	6.955	0.000	-3.011	8.782

SD: Standard deviation, KS: Kolmogorov-Smirnov-test

Details regarding the relationships between demographic characteristics of the management team and firm performance are displayed in the appendix. Team size and gender distribution were found to be statistically significantly related to firm performance. The percentage of males on the team has a positive impact on both growth ($\beta = 0.19$, $p < 0.05$) and profitability ($\beta = 0.19$, $p < 0.1$), although the latter effect is only marginally significant. Team size has a significant impact on growth ($\beta = 0.19$, $p < 0.05$).

7.1.7   Overview of executive team characteristics and firm performance

The analysis considered multiple aspects of the characteristics and composition of the executive team in research-based spin-off companies and examined the impacts of prior professional experience, educational background, team composition, heterogeneity, and demographic attributes and general characteristics on firm performance. In this context, the dimensions of growth and profitability were proven to be distinct facets of firm performance. Table 7.21 gives an overview of the results. Statistically significant links to performance were found only for prior management and industry work experience, diversity of team composition and the gender distribution in the team. This is an interesting finding in light of the extensive discussion about the importance of human capital, especially in the context of new technology-based ventures. Because fewer direct effects than expected were identified, the results raise the question of whether the characteristics of the management team influence performance outcomes indirectly.

Table 7.21: Overview of executive team characteristics and firm performance

		GROWTH	PROFIT
Prior professional experience	Entrepreneurial experience 1	$+^{\dagger}$	n.s.
	Management experience 1	$+^{*}$	n.s.
	Industry work experience 1	$+^{\dagger}$	$+^{*}$
Educational background	Management education	n.s.	n.s.
	Technology education	n.s.	n.s.
	Level of education	n.s.	n.s.
Heterogeneity	Diversity experience	n.s.	n.s.
	Diversity education	n.s.	n.s.
	Diversity composition	$+^{**}$	n.s.
Composition	External manager	n.s.	n.s.
	Ownership	n.s.	n.s.
	Founder	n.s.	n.s.
Demographic attributes and general characteristics	Team size	n.s.	n.s.
	Average age	n.s.	n.s.
	Gender distribution	$+^{*}$	$+^{\dagger}$

$^{\dagger}p < 0.1$; $^{*}p < 0.05$; $^{**}p < 0.01$; $^{***}p < 0.001$

7.2   Link between strategic decision making and firm performance

The following section reports the results of the examined relationships between strategic decision making and firm performance. In this context, two main areas are covered: (1) formal planning, measured by the intensity of strategic and long-term planning activities in contrast to informal, intuitive approaches that focus on flexibility, experimentation, and partnering; and (2) level of entrepreneurial strategic posture, including the five dimensions of innovativeness, proactiveness, risk-taking, competitive aggressiveness, and autonomy. First, the analysis gives an overview of the descriptive statistics and fit indices for the measurement models. Second, bivariate and multivariate analyses examine the impact on firm performance.

7.2.1   Strategic planning and firm performance

As outlined above, the latent construct measuring the intensity of strategic planning activities was operationalised by four items. The descriptive analysis displayed in table 7.22 shows the mean, standard deviation, skewness, and kurtosis for each indicator. The mean values for the strategic planning indicators indicate that spin-off companies tend to emphasise planning activities. In this respect, the second indicator seems to be most important for the respondents because it has a relatively high mean value. This indicator reflects whether the firm has a clear and consistent vision of its future direction. According to the Kolmogorov-Smirnov test, the indicators are not exactly normally distributed, but the overall mean of all indicators is not significant in terms of the null hypothesis and seems to be normally distributed. Furthermore, the absolute values of the skewness and kurtosis do not exceed 0.942 and 0.680, respectively, both of which are satisfactory.

Table 7.22: Descriptive statistics of latent variable planning

	N	Mean	SD	KS	P (KS)	Skewness	Kurtosis
PLAN 1	191	4.08	1.486	2.162	0.000	-0.097	-0.680
PLAN 2	193	5.34	1.421	3.450	0.000	-0.942	0.418
PLAN 3	192	4.77	1.552	2.789	0.000	-0.563	-0.603
PLAN 4	192	4.60	1.411	2.834	0.000	-0.356	-0.511
PLAN all	189	4.70	1.009	0.863	0.446	-0.282	-0.393

SD: Standard deviation, KS: Kolmogorov-Smirnov-test

In terms of convergent validity, table 7.23 presents the relevant fit indices for the measurement model, which are mostly in an acceptable range. The Cronbach's alpha value of 0.763 and item-to-total correlations above 0.508 suggest a sufficient level of internal consistency. The results of the explorative factor analysis indicate a satisfactory convergent validity, with factor loadings of at least 0.716 and an explained variance of 58.6 per cent. The confirmatory factor analysis reports that the first indicator is slightly below the required benchmark and could be eliminated from the scale. This item reflects the extent to which formal plans serve as a basis for competitive analyses. The overall measurement model should indicate the level of strategic planning activities, and formal plans can be seen as a crucial element. Thus, the indicator remains in the model. The reported factor reliability of 0.601 is satisfactory, whereas the AVE of 0.458 is slightly below the recommended benchmark.

Table 7.23: Fit indices for latent variable planning

Analysis	Fit index	Result
Internal consistency	Cronbach´s alpha (≥ 0.7)	0.763
	Item-to-total (≥ 0.5) PLAN 1 PLAN 2 PLAN 3 PLAN 4	0.508 0.539 0.597 0.610
Explorative factor analysis	Explained variance (≥ 0.5)	0.586
	Factor loading (≥ 0.5) PLAN 1 PLAN 2 PLAN 3 PLAN 4	0.716 0.747 0.794 0.803
Confirmatory factor analysis	Factor reliability (≥ 0.6)	0.601
	AVE (≥ 0.5)	0.458
	Indicator reliability (≥ 0.4) PLAN 1 PLAN 2 PLAN 3 PLAN 4	0.342 0.406 0.525 0.560

N = 191

According to hypothesis 2.1 the intensity of strategic planning activities is expected to have a positive impact on firm performance in the context of high-technology research-based spin-off companies. Table 7.24 reports the results regarding the impact on the performance dimensions of growth and profitability. The intensity of strategic planning activities has a statistically significant impact on growth ($\beta = 0.371$, $p < 0.001$), with a medium effect size and an explained variance of 13.8 per cent. However, the effect on profitability is only marginally significant ($\beta = 0.158$, $p < 0.1$), with a small effect size, and is responsible for only 2.5 per cent of the variance of the dependent variable. However, both effects have the same direction and confirm the predicted relationships. Because the fit indices fulfil the required criteria, the total fit of both models is good.

Table 7.24: Planning and firm performance

	GROWTH	PROFIT
Effect size	0.371^{***}	$0.158^{\dagger}$
R^2	0.138	0.025
Model fit:		
χ^2	14.1	10.6
df	13	8
P (≥ 0.1)	0.370	0.227
χ^2/df (≤ 3.0)	1.081	1.321
RMSEA (≤ 0.1)	0.021	0.041
CFI (≥ 0.9)	0.998	0.991
NFI (≥ 0.9)	0.972	0.965
PCLOSE (≥ 0.1)	0.751	0.533

$^{\dagger}p < 0.1$; $^*p < 0.05$; $^{**}p < 0.01$; $^{***}p < 0.001$

7.2.2 Futurity and firm performance

The futurity variable captures the firm-level attitude towards a long-term versus a short-term orientation of the company. The latent construct was originally composed of four reflective items. The first item was reverse coded before being used in the analysis. The descriptive statistics show that the means and standard deviations of the items are very similar. The Kolmogorov-Smirnov test makes clear that the indicators do not follow a normal distribution exactly, but the sum of all indicators does not statistically significantly ($p < 0.05$) reject the normal distribution assumption. In addition, the absolute value of the skewness of all indicators is below 1.021, and the maximum kurtosis score is 0.863. Both of these values are in an acceptable range in terms of a moderate deviation from the normality assumption. Table 7.25 provides the details.

Table 7.25: Descriptive statistics of latent variable futurity

	N	Mean	SD	KS	P (KS)	Skewness	Kurtosis
FUT 1	191	4.09	1.579	2.184	0.000	-1.021	0.350
FUT 2	193	4.26	1.580	2.615	0.000	-0.451	-0.706
FUT 3	192	4.32	1.490	2.789	0.000	-0.472	-0.515
FUT 4	192	4.42	1.530	2.618	0.000	-0.259	-0.863
FUT all	191	4.27	1.075	1.269	0.080	-0.460	-0.155

SD: Standard deviation, KS: Kolmogorov-Smirnov-test

The first item was eliminated from the construct because it had a very low item-to-total correlation of 0.119. The latent variable with the remaining three indicators seems to have sufficient internal consistency, with Cronbach's alpha of 0.784 and item-to-total correlations above 0.502. The explorative factor analysis provides acceptable results, with an explained variance of 70.1 per cent and factor loadings of at least 0.742. The confirmatory factor analysis yields satisfactory factor reliability of 0.800 and AVE of 0.579. Nevertheless, the fourth indicator is below the benchmark in terms of indicator reliability at only 0.305, and thus it could presumably be removed from the scale. Because the overall fit of the measurement model is acceptable and the indicators were drawn from scales established in previous research (Venkatraman, 1989), this item remains in the model. The details of the discussed fit indices are displayed in table 7.26.

Table 7.26: Fit indices for latent variable futurity

Analysis	Fit index	Result
Internal consistency	Cronbach´s alpha (≥ 0.7)	0.784
	Item-to-total (≥ 0.5) FUT 1 FUT 2 FUT 3 FUT 4	- 0.676 0.701 0.502
Explorative factor analysis	Explained variance (≥ 0.5)	0.701
	Factor loading (≥ 0.5) FUT 1 FUT 2 FUT 3 FUT 4	- 0.876 0.887 0.742
Confirmatory factor analysis	Factor reliability (≥ 0.6)	0.800
	AVE (≥ 0.5)	0.579
	Indicator reliability (≥ 0.4) FUT 1 FUT 2 FUT 3 FUT 4	- 0.679 0.756 0.305

N = 192

According to hypothesis 2.2, a long-term perspective of the company, reflected by the latent variable futurity, should be positively related to firm performance. In fact, the results support this proposition for both the growth and profitability dimensions of performance. In the context of academic spin-offs, futurity has a significant positive effect on growth (β = 0.337, $p < 0.001$), with a medium effect size, explaining 11.3 per cent of the variance of the dependent variable. Long-term planning also positively and significantly influences the profitability of the ventures (β = 0.166, $p < 0.05$). The effect size is medium, and the variable explains 2.7 per cent of the variance in the dependent variable. Table 7.27 presents the results regarding the impact of futurity on the performance dimensions of growth and profitability. The total fit of the models is tolerable because almost all criteria are in an acceptable range.

Table 7.27: Futurity and firm performance

	GROWTH	PROFIT
Effect size	0.337***	0.166*
R^2	0.113	0.027
Model fit:		
χ^2	11.4	9.1
df	8	4
P ($\geq$ 0.1)	0.178	0.060
χ^2/df ($\leq$ 3.0)	1.429	2.265
RMSEA ($\leq$ 0.1)	0.047	0.081
CFI ($\geq$ 0.9)	0.993	0.983
NFI ($\geq$ 0.9)	0.977	0.971
PCLOSE ($\geq$ 0.1)	0.469	0.187

†$p < 0.1$; *$p < 0.05$; **$p < 0.01$; ***$p < 0.001$

7.2.3 Flexibility and firm performance

In contrast to the previous two constructs, which captured aspects such as formal planning and long-term forecasting, the following variables reflect perspectives towards more informal, intuitive, and incremental approaches in the context of strategic decision making. The next section analyses whether flexibility, experimentation, and partnering have a significant impact on firm performance. First, as outlined in chapter 6, the original operationalisation of the flexibility construct consisted of four items. According to the descriptive analysis, the means of the indicators are relatively high, ranging between 4.90 and 5.75. The scores of the single indicators and the overall arithmetic average are not normally distributed according to Kolmogorov-Smirnov significance test ($p < 0.05$). The absolute values for the skewness and kurtosis of the third indicator score highest, at 1.300 and 2.845, respectively. This indicator captures the extent to which firms flexibly take advantage of opportunities. According to Kline (2005), both outcomes are tolerable, so the assumption of a normal distribution is only moderately violated. Table 7.28 presents the descriptive statistics of the indicators.

Table 7.28: Descriptive statistics of latent variable flexibility

	N	Mean	SD	KS	P (KS)	Skewness	Kurtosis
FLEX 1	191	5.26	1.278	2.989	0.000	-0.966	1.251
FLEX 2	192	5.01	1.481	2.651	0.000	-0.673	-0.250
FLEX 3	189	5.75	1.143	3.136	0.000	-1.300	2.845
FLEX 4	191	4.90	1.356	2.756	0.000	-0.548	0.350
FLEX all	188	5.23	0.943	1.428	0.034	-1.012	2.741

SD: Standard deviation, KS: Kolmogorov-Smirnov-test

The second and the fourth items of the latent variable were eliminated because fit indices for the measurement model reported poor item-to-total correlations of 0.439 and 0.325, respectively. Table 7.29 shows that the fit of the measurement model consisting of the two remaining indicators is in an acceptable range for all reported indices. Cronbach's alpha of 0.744 and the item-to-total correlations of 0.596 indicate adequate internal consistency. Explorative and confirmatory factor analysis yield satisfactory results, indicating sufficient convergent validity of the latent variable flexibility.

Table 7.29: Fit indices for latent variable flexibility

Analysis	Fit index	Result
Internal consistency	Cronbach´s alpha ($\geq$ 0.7)	0.744
	Item-to-total ($\geq$ 0.5) FLEX 1 FLEX 2 FLEX 3 FLEX 4	0.596 - 0.596 -
Explorative factor analysis	Explained variance ($\geq$ 0.5)	0.798
	Factor loading ($\geq$ 0.5) FLEX 1 FLEX 2 FLEX 3 FLEX 4	0.893 - 0.893 -
Confirmatory factor analysis	Factor reliability ($\geq$ 0.6)	0.748
	AVE ($\geq$ 0.5)	0.598
	Indicator reliability ($\geq$ 0.4) FLEX 1 FLEX 2 FLEX 3 FLEX 4	0.606 - 0.587 -

N = 189

In chapter 4, the relationship between a flexible approach in terms of strategic decision making and firm performance was addressed by a research question because neither theory nor empirical findings could support a robust assumption for the direction of a potential impact. Thus, this part of the analysis has an explorative character. As discussed in the previous chapters, flexibility could be important in the context of research-based and technology-oriented ventures, where identifying, exploring, and exploiting opportunities play a major role. However, the results displayed in table 7.30 do not support these considerations. No statistically significant relationship could be found. In view of the goodness-of-fit indicators, none of the models seem to be problematic.

Table 7.30: Flexibility and firm performance

	GROWTH	PROFIT
Effect size	0.166	0.178
R^2	0.028	0.032
Model fit:		
χ^2	3.5	0.027
df	4	1
P (≥ 0.1)	0.483	0.870
χ^2/df (≤ 3.0)	0.867	0.027
RMSEA (≤ 0.1)	0.000	0.000
CFI (≥ 0.9)	1.000	1.000
NFI (≥ 0.9)	0.991	1.000
PCLOSE (≥ 0.1)	0.690	0.898

$^\dagger p < 0.1$; $^*p < 0.05$; $^{**}p < 0.01$; $^{***}p < 0.001$

7.2.4 Experimentation and firm performance

Besides the flexibility construct outlined in the previous section, the next latent variable is supposed to capture the level of experimentation in the company. The original measurement model consists of two items. Descriptive statistics for the items are displayed in table 7.31 and show large differences in the means of the indicators. Furthermore, the standard deviations are very different at 3.68 and 4.73, respectively. The Kolmogorov-Smirnov test shows that the indicators are not perfectly normally distributed, although the sum of the indicators is not statistically significant in terms of the null hypothesis ($p < 0.05$). The skewness and kurtosis are in an acceptable range. Absolute values are not higher than 0.666 and 1.250, respectively.

Table 7.31: Descriptive statistics of latent variable experimentation

	N	Mean	SD	KS	P (KS)	Skewness	Kurtosis
EXP 1	191	4.73	1.666	3.159	0.000	-0.666	-0.529
EXP 2	192	3.68	1.960	2.623	0.000	0.334	-1.250
EXP all	190	4.22	1.394	1.281	0.075	-0.085	-0.669

SD: Standard deviation, KS: Kolmogorov-Smirnov-test

Table 7.32 clearly shows that several criteria demonstrate a poor fit to the measurement model. Cronbach's alpha of 0.307 and item-to-total correlations of 0.184 both indicate insufficient internal consistency. Although the results for the explorative factor analysis are in an acceptable range, the confirmatory factor analysis yields low scores for factor reliability and AVE. The second indicator has a very low indicator reliability of only 0.037. Therefore, this measurement model cannot be used for further analyses and has to be excluded.

Table 7.32: Fit indices for latent variable experimentation

Analysis	Fit index	Result
Internal consistency	Cronbach's alpha (≥ 0.7)	0.307
	Item-to-total (≥ 0.5) EXP 1 EXP 2	0.184 0.184
Explorative factor analysis	Explained variance (≥ 0.5)	0.592
	Factor loading (≥ 0.5) EXP 1 EXP 2	0.925 0.037
Confirmatory factor analysis	Factor reliability (≥ 0.6)	0.503
	AVE (≥ 0.5)	0.410
	Indicator reliability (≥ 0.4) EXP 1 EXP 2	0.925 0.037

N = 189

7.2.5 Partnering and firm performance

The latent variable partnering, which reflects the level of informal collaboration with partners, is specified by two items. The descriptive overview shown in table 7.33 demonstrates that the mean values of both indicators are relatively high, at 4.82 and 5.16, respectively. Informal partnerships are likely to be an important aspect of research-based spin-off companies. The

Kolmogorov-Smirnov tests reveal problematic findings for the shape of the distribution on the indicator level and for the overall arithmetic mean value. However, skewness and kurtosis are both in an acceptable range because the absolute values do not exceed 0.950 for skewness and 0.572 for kurtosis.

Table 7.33: Descriptive statistics of latent variable partnering

	N	Mean	SD	KS	P (KS)	Skewness	Kurtosis
PART 1	191	4.82	1.689	2.715	0.000	-0.651	-0.572
PART 2	191	5.16	1.509	3.021	0.000	-0.950	0.405
PART all	191	4.99	1.481	2.180	0.000	-0.745	-0.153

SD: Standard deviation, KS: Kolmogorov-Smirnov-test

Table 7.34 presents the scores reflecting the fit of the measurement model. All of the discussed criteria are in a satisfactory range. Internal consistency is demonstrated by a Cronbach's alpha score of 0.832 and item-to-total correlation of 0.716. The explorative factor analysis also leads to good results, with an explained variance of 85.8 per cent and factor loadings of 0.926. The confirmatory factor analysis delivers satisfactory outcomes as well, with a factor reliability of 0.842 and AVE of 0.730.

Table 7.34: Fit indices for latent variable partnering

Analysis	Fit index	Result
Internal consistency	Cronbach´s alpha ($\geq$ 0.7)	0.832
	Item-to-total ($\geq$ 0.5) PART 1 PART 2	0.716 0.716
Explorative factor analysis	Explained variance ($\geq$ 0.5)	0.858
	Factor loading ($\geq$ 0.5) PART 1 PART 2	0.926 0.926
Confirmatory factor analysis	Factor reliability ($\geq$ 0.6)	0.842
	AVE ($\geq$ 0.5)	0.730
	Indicator reliability ($\geq$ 0.4) PART 1 PART 2	0.791 0.649

N = 189

In line with the discussion about the potential impact of flexibility on performance, the relationship between a strong emphasis on informal partnerships and collaborations and firm performance is addressed by a research question because previous research has not provided a grounded theoretical explanation or empirical evidence. However, exploring this link in the context of research-based spin-off companies seems to be promising. Table 7.35 reports the results regarding the impact on the performance dimensions of growth and profitability. Partnering has a statistically significant impact on growth (β = 0.234, p < 0.01) and profitability (β = 0.234, p < 0.05). This factor explains 5.5 per cent of the variance of both dependent variables. The fit indices indicate proper model fits.

Table 7.35: Partnering and firm performance

	GROWTH	PROFIT
Effect size	0.234[**]	0.234[*]
R^2	0.055	0.055
Model fit:		
χ^2	4.6	1.2
df	4	1
P ($\geq$ 0.1)	0.334	0.282
χ^2/df ($\leq$ 3.0)	1.143	1.158
RMSEA ($\leq$ 0.1)	0.027	0.029
CFI ($\geq$ 0.9)	0.999	0.999
NFI ($\geq$ 0.9)	0.990	0.995
PCLOSE ($\geq$ 0.1)	0.558	0.389

[†]p < 0.1; [*]p < 0.05; [**]p < 0.01; [***]p < 0.001

7.2.6   Innovativeness and firm performance

The innovativeness construct, describing a strategic orientation with an emphasis on novelty and change, was originally operationalised by three items. Descriptive statistics of the indicators are displayed in table 7.36. The first and second indicators in particular demonstrate high mean values of 5.63 and 4.60, respectively. These indicators reflect a strong emphasis on R&D, innovation, and the introduction of new products or services. Thus, high scores are not surprising in the context of new research- and technology-based ventures. The second factor shows relatively high skewness and kurtosis, with absolute values of 1.459 and 2.250, respectively. The Kolmogorov-Smirnov test does not support the normality assumption with regard to the distribution of the indicators and the overall average, but the maximum absolute values of 1.459 for skewness and 2.250 for kurtosis are still tolerable (Kline, 2005). However, the model fit might be influenced negatively.

Table 7.36: Descriptive statistics of latent variable innovativeness

	N	Mean	SD	KS	P (KS)	Skewness	Kurtosis
INNO 1	193	5.62	1.372	3.364	0.000	-1.077	0.678
INNO 2	193	5.53	1.369	3.661	0.000	-1.459	2.250
INNO 3	192	4.60	1.388	2.566	0.000	-0.580	-0.003
INNO all	191	5.26	0.993	1.970	0.001	-0.769	0.135

SD: Standard deviation, KS: Kolmogorov-Smirnov-test

The values in table 7.37 make clear that the fit of the measurement model does not meet the necessary requirements for some of the criteria. The second item was eliminated because it had the lowest item-to-total correlation of only 0.302. Nevertheless, the construct defined by the two remaining items still seems to be problematic in terms of model fit on the measurement level. Cronbach's alpha is 0.515, which is still acceptable for measurement models with less than four items (Peter, 1997). The item-to-total correlation is only 0.347, indicating a low level of internal consistency of the construct. However, the explorative factor

Table 7.37: Fit indices for latent variable innovativeness

Analysis	Fit index	Result
Internal consistency	Cronbach's alpha (≥ 0.7)	0.515
	Item-to-total (≥ 0.5) INNO 1 INNO 2 INNO 3	- 0.347 0.347
Explorative factor analysis	Explained variance (≥ 0.5)	0.673
	Factor loading (≥ 0.5) INNO 1 INNO 2 INNO 3	- 0.821 0.821
Confirmatory factor analysis	Factor reliability (≥ 0.6)	0.515
	AVE (≥ 0.5)	0.347
	Indicator reliability (≥ 0.4) INNO 1 INNO 2 INNO 3	- 0.340 0.355

N = 192

analysis leads to satisfactory results, with an explained variance of 67.3 per cent and factor loadings of 0.821. The confirmatory factor analysis confirms the poor model fit, with a factor reliability of 0.515 and AVE of 0.347. The indicators probably did not measure the same phenomenon. Whereas item 2 asked for the number of new products and services that have been introduced recently, item 3 addresses the extent of changes in the portfolio. The construct will not be excluded from further analysis, but the findings need to be discussed in light of the disclosed constraints in terms of the fit of the measurement model.

In hypothesis 2.4.1, the level of innovativeness was expected to have a positive impact on firm performance. Because the fit of the measurement model is weak, the results have to be discussed with caution. However, table 7.38 reports a statistically significant positive impact on growth ($\beta = 0.398$, $p < 0.001$), with a medium effect size and an explained variance of 15.8 per cent. The link between innovativeness and profitability was found to be marginally significant ($\beta = 0.279$, $p < 0.1$), also with a medium effect size. In this model, innovativeness is responsible for 7.8 per cent of the variance of the dependent variable of profitability. The tests for the overall model fit yielded positive results.

Table 7.38: Innovativeness and firm performance

	GROWTH	PROFIT
Effect size	0.398[**]	0.279[†]
R^2	0.158	0.078
Model fit:		
χ^2	2.8	0.3
df	4	1
P (≥ 0.1)	0.587	0.596
χ^2/df (≤ 3.0)	0.707	0.280
RMSEA (≤ 0.1)	0.000	0.000
CFI (≥ 0.9)	1.000	1.000
NFI (≥ 0.9)	0.991	0.998
PCLOSE (≥ 0.1)	0.769	0.676

[†]$p < 0.1$; [*]$p < 0.05$; [**]$p < 0.01$; [***]$p < 0.001$

7.2.7 Proactiveness and firm performance

In chapter 6, the variable proactiveness was defined as a distinct dimension of the broader concept of entrepreneurial orientation and reflected by three indicators. Referring to the descriptive overview of the data that is presented in table 7.39, the mean value of the second indicator scores relatively high, at 5.50. This item captures whether the firm is very often or very seldom the first to introduce new products or services. Thus, this aspect seems to be crucial for research-based spin-off companies. The Kolmogorov-Smirnov test does not confirm a normal distribution for the indicators or for the overall value. The absolute values of skewness and kurtosis do not exceed their respective maxima of 1.343 and 2.079 and are still in an acceptable range.

Table 7.39: Descriptive statistics of latent variable proactiveness

	N	Mean	SD	KS	P (KS)	Skewness	Kurtosis
PRO 1	193	4.89	1.295	2.618	0.000	-0.369	0.107
PRO 2	193	5.50	1.447	3.506	0.000	-1.343	1.596
PRO 3	192	4.58	1.300	3.567	0.000	-1.329	2.079
PRO all	192	5.33	1.064	1.831	0.002	-0.913	1.503

SD: Standard deviation, KS: Kolmogorov-Smirnov-test

As reported in table 7.40, some criteria regarding the fit of the measurement model are not in line with recommended thresholds. Although Cronbach's alpha coefficient is sufficient at 0.704, the first indicator lies slightly below the recommended threshold in terms of item-to-total correlation and scores only 0.483. This indicator reflects whether the spin-off company primarily responds to actions by its competitors or initiates actions itself. The explorative factor analysis leads to satisfactory results for all of the indicators, with factor loadings above 0.761 and an explained variance of 62.9 per cent. The confirmatory factor analysis also demonstrates that the first indicator does not perfectly match the construct. The indicator reliability is only 0.353, leading to an AVE of 0.455, which is also a bit below the recommended score. However, the factor reliability of 0.712 is good. Therefore, the indicator remains in the model because it covers an important aspect of the construct and the overall fit remains tolerable.

Table 7.40: Fit indices for latent variable proactiveness

Analysis	Fit index	Result
Internal consistency	Cronbach´s alpha (≥ 0.7)	0.704
	Item-to-total (≥ 0.5) PRO 1 PRO 2 PRO 3	0.483 0.546 0.537
Explorative factor analysis	Explained variance (≥ 0.5)	0.629
	Factor loading (≥ 0.5) PRO 1 PRO 2 PRO 3	0.761 0.812 0.804
Confirmatory factor analysis	Factor reliability (≥ 0.6)	0.712
	AVE (≥ 0.5)	0.455
	Indicator reliability (≥ 0.4) PRO 1 PRO 2 PRO 3	0.353 0.513 0.481

N = 192

Hypothesis 2.4.2 proposed a positive relationship between proactiveness and firm performance. Corresponding to the results reported in table 7.41, proactiveness has a statistically significant positive impact on firm growth ($\beta = 0.400$, $p < 0.001$), with a medium effect size and an explained variance of 16.0 per cent of the dependent variable. On the other hand, the relationship between proactiveness and profitability was not found to be statistically significant. In comparison with other factors, proactiveness as a single factor accounts for a relatively large effect size and a large proportion of explained variance. The total fit of the model is tolerable, although the probability value of the chi-squared test of goodness-of-fit is not reached, probably due to the sensitivity of the test for large samples.

Table 7.41: Proactiveness and firm performance

	GROWTH	PROFIT
Effect size	0.400***	0.187
R^2	0.160	0.035
Model fit:		
χ^2	16.3	8.1
df	8	4
P ($\geq$ 0.1)	0.038	0.087
χ^2/df ($\leq$ 3.0)	2.043	2.030
RMSEA ($\leq$ 0.1)	0.074	0.073
CFI ($\geq$ 0.9)	0.979	0.980
NFI ($\geq$ 0.9)	0.962	0.963
PCLOSE ($\geq$ 0.1)	0.193	0.240

†p < 0.1; *p < 0.05; **p < 0.01; ***p < 0.001

7.2.8 Risk-taking and firm performance

The risk-taking dimension was introduced as a latent variable with three reflective indicators. The third item was originally defined as an inverse measure, so it had to be reverse coded before being used in the analysis. The Kolmogorov-Smirnov test shows that the indicators and the overall average do not perfectly follow a normal distribution. However, both skewness and kurtosis are in an acceptable range and do not exceed absolute values of 0.313 and 0.661, respectively. More details about the descriptive statistics of the indicators are given in table 7.42.

Table 7.42: Descriptive statistics of latent variable risk-taking

	N	Mean	SD	KS	P (KS)	Skewness	Kurtosis
RISK 1	193	4.51	1.511	2.073	0.000	-0.313	-0.661
RISK 2	191	3.96	1.331	2.214	0.000	-0.095	-0.613
RISK 3	191	4.30	1.197	2.492	0.000	-0.271	-0.437
RISK all	189	4.27	0.998	1.482	0.025	0.032	0.022

SD: Standard deviation, KS: Kolmogorov-Smirnov-test

In table 7.43, the fit indices for the measurement model are displayed. Cronbach's alpha is only 0.588, which is still acceptable for a model with fewer than 4 items (Peter, 1997). However, the item-to total correlations are all below the recommended threshold, indicating a poor internal consistency of the indicators. Excluding one of the items from the measurement model did not improve the overall fit. Explorative factor analysis indicates sufficient outcomes, with explained variance of 54.9 per cent and factor loadings of at least 0.705. In the confirmatory factor analysis, satisfactory factor reliability of 0.611 was obtained. However, the indicator reliabilities of the second and third items are below the recommended benchmark, and AVE is only 0.353. In sum, there are mixed results for the fit of the measurement model. Thus, the construct will still be used in further analysis, but the potential results need be discussed with caution.

Table 7.43: Fit indices for latent variable risk-taking

Analysis	Fit index	Result
Internal consistency	Cronbach´s alpha ($\geq$ 0.7) .	0.588
	Item-to-total ($\geq$ 0.5) RISK 1 RISK 2 RISK 3	0.430 0.406 0.365
Explorative factor analysis	Explained variance ($\geq$ 0.5)	0.549
	Factor loading ($\geq$ 0.5) RISK 1 RISK 2 RISK 3	0.771 0.747 0.705
Confirmatory factor analysis	Factor reliability ($\geq$ 0.6)	0.611
	AVE ($\geq$ 0.5)	0.353
	Indicator reliability ($\geq$ 0.4) RISK 1 RISK 2 RISK 3	0.433 0.335 0.259

N = 192

According to hypothesis 2.43, a risky approach should have a positive impact on firm performance, especially in the context of new technology-based ventures. Table 7.44 shows the results regarding the link between risk-taking and the performance dimensions of growth and profitability. No significant results could be identified. The total fit of the model is appropriate.

Table 7.44: Risk-taking and firm performance

	GROWTH	PROFIT
Effect size	0.129	0.022
R^2	0.017	0.000
Model fit:		
χ^2	4.2	7.4
df	8	4
P ($\geq$ 0.1)	0.842	0.114
χ^2/df ($\leq$ 3.0)	0.520	1.861
RMSEA ($\leq$ 0.1)	0.000	0.067
CFI ($\geq$ 0.9)	1.000	0.978
NFI ($\geq$ 0.9)	0.988	0.957
PCLOSE ($\geq$ 0.1)	0.954	0.286

$^\dagger p < 0.1$; $^*p < 0.05$; $^{**}p < 0.01$; $^{***}p < 0.001$

7.2.9 Competitive aggressiveness and firm performance

The latent variable competitive aggressiveness was conceptualised by two reflective indicators. The second item was originally defined as an inverse measure and reverse coded for use in the analysis. Similar to some constructs in previous sections, the Kolmogorov-Smirnov test does not demonstrate an exact normal distribution of the indicators or the overall average. Nevertheless, skewness and kurtosis are both in an adequate range. Table 7.45 reports maximum absolute values of 0.196 and 0.713 for skewness and kurtosis, respectively.

Table 7.45: Descriptive statistics of latent variable competitive aggressiveness

	N	Mean	SD	KS	P (KS)	Skewness	Kurtosis
AGG 1	191	4.10	1.561	2.290	0.000	-0.142	-0.713
AGG 2	191	3.45	1.450	2.078	0.000	0.196	-0.662
AGG all	190	3.79	1.296	1.615	0.011	-0.100	-0.474

SD: Standard deviation, KS: Kolmogorov-Smirnov-test

As displayed in table 7.46, most of the fit indices with regard to the measurement model are in a satisfactory range. The Cronbach's alpha of 0.667 is sufficient according to Peter (1997) because the measurement model only consists of two items. The item-to-total correlations of 0.502 are slightly above the benchmark. The explorative factor analysis reports an explained variance of 71.2 per cent and factor loadings of 0.866. The confirmatory factor analysis leads to one critical outcome. The indicator reliability of the second item is only 0.332, which is below the recommended score. However, the factor reliability of 0.712 and AVE of 0.567 indicate an appropriate model fit. Thus, the model will be used in the further analysis.

Table 7.46: Fit indices for latent variable competitive aggressiveness

Analysis	Fit index	Result
Internal consistency	Cronbach´s alpha ($\geq$ 0.7)	0.667
	Item-to-total ($\geq$ 0.5) AGG 1 AGG 2	 0.502 0.502
Explorative factor analysis	Explained variance ($\geq$ 0.5)	0.751
	Factor loading ($\geq$ 0.5) AGG 1 AGG 2	 0.866 0.866
Confirmatory factor analysis	Factor reliability ($\geq$ 0.6)	0.712
	AVE ($\geq$ 0.5)	0.567
	Indicator reliability ($\geq$ 0.4) AGG 1 AGG 2	 0.771 0.332

N = 192

Competitive aggressiveness should lead to higher performance, as suggested by hypothesis 2.4.4. The outcomes reveal a statistically significant positive effect of competitive aggressiveness on firm growth ($\beta = 0.307$, $p < 0.05$) with a medium effect size. This factor explains 9.4 per cent of the variance. Competitive aggressiveness and profitability were not significantly related. The total fits of the models are good.

Table 7.47: Competitive aggressiveness and firm performance

	GROWTH	PROFIT
Effect size	0.307[*]	0.111
R^2	0.094	0.019
Model fit:		
χ^2	6.4	0.2
df	4	1
P ($\geq$ 0.1)	0.170	0.691
χ^2/df ($\leq$ 3.0)	1.603	0.158
RMSEA ($\leq$ 0.1)	0.056	0.000
CFI ($\geq$ 0.9)	0.993	1.000
NFI ($\geq$ 0.9)	0.982	0.999
PCLOSE ($\geq$ 0.1)	0.369	0.754

[†]$p < 0.1$; [*]$p < 0.05$; [**]$p < 0.01$; [***]$p < 0.001$

7.2.10 Autonomy and firm performance

The autonomy construct originally consisted of three items. The first item was defined as an inverse measure and recoded before running the analysis. Descriptive statistics are presented in table 7.48. The Kolmogorov-Smirnov test does not confirm an exact normal distribution on the indicator level, but the overall average is not statistically significant in terms of the null hypothesis (p < 0.05). Additionally, the absolute value for skewness is 1.038 and for kurtosis is 1.372, so the discrepancy between the observed distribution and the assumed normal distribution is moderate.

Table 7.48: Descriptive statistics of latent variable autonomy

	N	Mean	SD	KS	P (KS)	Skewness	Kurtosis
AUT 1	192	5.18	1.399	3.722	0.000	-1.038	0.597
AUT 2	192	4.02	1.899	2.279	0.000	0.010	-1.372
AUT 3	191	5.38	1.275	2.533	0,000	-0.918	1.331
AUT all	190	4.853	1.115	1.028	0.241	-0.242	-0.236

SD: Standard deviation, KS: Kolmogorov-Smirnov-test

Because of a low item-to-total correlation, the first indicator was eliminated from the scale. The remaining two items report a Cronbach's alpha of 0.551, which does not fulfil the discussed benchmark but is tolerable because the construct consists of only two items (Peter, 1997). The explorative factor analysis resulted in satisfactory outcomes with an explained variance of 70.5 per cent and factor loadings of 0.840. However, the confirmatory factor analysis reported factor reliability and AVE above 1, which are not in the designated range of between 0 and 1 for both indicators. The indicator reliability of the second item is 1.797, which is also outside the allowed interval. For this reason, the latent variable autonomy cannot be considered in the subsequent analysis. Table 7.49 gives a detailed overview of the fit indices.

Table 7.49: Fit indices for latent variable autonomy

Analysis	Fit index	Result
Internal consistency	Cronbach´s alpha (≥ 0.7)	0.551
	Item-to-total (≥ 0.5) AUT 1 AUT 2 AUT 3	 - 0.410 0.410
Explorative factor analysis	Explained variance (≥ 0.5)	0.705
	Factor loading (≥ 0.5) AUT 1 AUT 2 AUT 3	 - 0.840 0.840
Confirmatory factor analysis	Factor reliability (≥ 0.6)	1.195
	AVE (≥ 0.5)	1.267
	Indicator reliability (≥ 0.4) AUT 1 AUT 2 AUT 3	 - 1.797 0.094

N = 192

7.2.11 Overview of strategic decision making and firm performance

The previous sections first examined the impact of formal and informal planning activities on the performance dimensions of growth and profitability, then studied the link between an entrepreneurial strategic posture and firm performance. Table 7.50 presents the key findings. The measurement models for the latent variables of planning, futurity, flexibility, and partnering were found to be acceptable. The model capturing experimentation did not work properly and was not considered in the analysis. The constructs of planning, futurity, and partnering were found to be statistically significantly and positively associated with firm growth. Futurity and partnering each have a statistically significant impact on profitability as well, whereas the link between planning and profitability is only marginally significant. The measurement models for the entrepreneurial orientation dimensions of proactiveness and competitive aggressiveness are in an overall acceptable range, whereas the facets of innovativeness, risk-taking, and autonomy were critical in some aspects. Thus, the findings have to be discussed in light of these constraints. Innovativeness, proactiveness, and competitive aggressiveness were found to be statistically significantly and positively related to growth. Innovativeness is also marginally positively related to profitability. Risk-taking has no statistically significant impact on firm performance.

Table 7.50: Overview of strategic decision making and firm performance

		GROWTH	PROFIT
Formal and informal planning	Planning	+***	+†
	Futurity	+***	+*
	Flexibility	n.s.	n.s.
	Experimentation	excl.	excl.
	Partnering	+*	+*
Entrepreneurial strategic posture	Innovativeness	+**	+†
	Proactiveness	+***	n.s.
	Risk-taking	n.s.	n.s.
	Competitive aggressiveness	+*	n.s.
	Autonomy	excl.	excl.

†p < 0.1; *p < 0.05; **p < 0.01; ***p < 0.001

7.3   Link between executive characteristics and strategic decision making

The next section presents the results regarding the link between characteristics of the executive team and strategic decision making. Thus, various aspects of decision making are dependent variables, whereas team characteristics such as prior experience, educational background, composition, and diversity are explanatory variables. This chapter consists of two main subsections. First, formal strategy approaches to strategic decision making focussing on rationality and planning are contrasted with informal processes using intuitive and heuristic reasoning. Second, potential team level antecedents of an entrepreneurial strategic posture are discussed. Only selected results are presented in the following paragraphs; the remaining findings and null results can be found in the appendix.

7.3.1   Prior professional experience and strategic decision making

The results regarding the impact of prior professional experience on the intensity of strategic planning activities are shown in tables 7.51 and 7.52. Previous experience is assessed on the team level by following two different approaches. The variables of the first approach capture whether at least one member of the executive team had specific experience before joining the spin-off, whereas the measures of the second approach capture the proportion of managers with the discussed type of professional experience. In models 1 to 3 of each table, the effect is measured for every variable separately, whereas model 4 contains the full model including all variables regarding professional experience. Hypotheses 3.1.2 and 3.1.3 predicted that prior entrepreneurial, management, and industry work experience lowers the level of rational planning activities and increases the emphasis on intuitive and heuristic strategising. In fact, the opposite result is found to be statistically significant. In both approaches, prior entrepreneurial experience has a statistically significant impact on the intensity of strategic planning activities ($\beta_1 = 0.231$, $p < 0.01$; $\beta_2 = 0.181$, $p < 0.05$). In the first approach, the effect size is medium, and the explained variance is 5.3 per cent, whereas the second approach accounts only for 3.3 per cent of the explained variance, with a small effect size. In a similar way, prior management experience is associated with strategic planning ($\beta_1 = 0.238$, $p < 0.01$; $\beta_2 = 0.222$, $p < 0.01$). Both approaches demonstrate a medium effect size and explained variance of 5.7 and 4.9 per cent, respectively. Previous industry work experience also influences the level of strategic planning activities positively ($\beta_1 = 0.276$, $p < 0.01$; $\beta_2 = 0.189$, $p < 0.05$), explaining 7.6 per cent of the variance in the first approach and 3.6 per cent in the second one.

In sum, in the first approach to measuring prior experience on the team level with an indicator variable, effect sizes are all medium, and the explained variance for each single variable lies between 5.3 and 6.9 per cent. The full model explains 12.6 per cent of the total variance, but only prior industry work experience remains statistically significant ($\beta = 0.225$, $p < 0.01$), and prior entrepreneurial experience is marginally significant ($\beta = 0.164$, $p < 0.1$). The high correlations between the explanatory variables have to be taken into consideration. Most of the criteria in terms of model fit are fulfilled; only the probability value of the chi-squared test of goodness-of-fit is not reached in models 2 and 4. In the second approach, only the effect size of the prior management experience variable is medium, whereas the others are small. The explained variance of the dependent variable is between 3.3 and 4.9 per cent. In the full model, none of the variables remains statistically significant. The model fit is good.

Table 7.51: Prior professional experience 1 and planning

	Model 1	Model 2	Model 3	Model 4
Entrepreneurial experience 1	0.231**			0.164[†]
Management experience 1		0.238**		0.110
Industry work experience 1			0.276**	0.225**
R^2	0.053	0.057	0.076	0.126
χ^2	4.3	10.7	3.8	17.9
df	5	5	5	11
P (≥ 0.1)	0.505	0.057	0.572	0.083
χ^2/df (≤ 3.0)	0.862	2.144	0.769	1.631
RMSEA (≤ 0.1)	0.000	0.077	0.000	0.057
CFI (≥ 0.9)	1.000	0.969	1.000	0.971
NFI (≥ 0.9)	0.978	0.947	0.981	0.933
PCLOSE (≥ 0.1)	0.730	0.199	0.778	0.355

[†]$p < 0.1$; *$p < 0.05$; **$p < 0.01$; ***$p < 0.001$

Table 7.52: Prior professional experience 2 and planning

	Model 1	Model 2	Model 3	Model 4
Entrepreneurial experience 2	0.181*			0.135
Management experience 2		0.222**		0.140
Industry work experience 2			0.189*	0.132
R^2	0.033	0.049	0.036	0.080
χ^2	4.5	8.6	3.4	15.2
df	5	5	5	11
P ($\geq$ 0.1)	0.474	0.126	0.637	0.172
χ^2/df ($\leq$ 3.0)	0.909	1.720	0.682	1.385
RMSEA ($\leq$ 0.1)	0.000	0.061	0.000	0.045
CFI ($\geq$ 0.9)	1.000	0.980	1.000	0.981
NFI ($\geq$ 0.9)	0.976	0.957	0.982	0.939
PCLOSE ($\geq$ 0.1)	0.706	0.328	0.822	0.514

$^{\dagger}p < 0.1$; $^{*}p < 0.05$; $^{**}p < 0.01$; $^{***}p < 0.001$

The next step of the analysis covers the long-term orientation of the company, measured by the latent variable futurity. As outlined in the previous section, prior professional experience is measured in two different ways. According to the results displayed in tables 7.53 and 7.54, previous entrepreneurial and management experience has a statistically significant impact on futurity. In this context, prior management experience shows a medium effect size in both approaches ($\beta_1 = 0.264$, $p < 0.001$; $\beta_2 = 0.207$, $p < 0.01$) and explains 7.0 and 4.3 per cent of the variance, respectively. On the other hand, entrepreneurial experience shows a medium effect size in the first approach ($\beta_1 = 0.227$, $p < 0.01$), with an explained variance of 5.1 per cent, and a small effect in the second ($\beta_2 = 0.167$, $p < 0.05$), with an explained variance of 2.8 per cent. Previous industry work experience was found to have only a marginally significant effect in the first approach ($\beta_1 = 0.150$, $p < 0.01$). No statistically significant relationship could be identified in the second approach.

The full model of the first approach shows that only management experience remains statistically significant ($\beta = 0.186$, $p < 0.05$), whereas entrepreneurial experience becomes marginally significant ($\beta = 0.149$, $p < 0.1$). In this model, 9.4 per cent of the total variance is explained, whereas single variables are responsible for 2.3 to 7.0 per cent of the variance. The overall model fit seems to be good because none of the discussed indices is violated. The full model of the second approach explains 5.9 per cent of the total variance, with single variables contributing between 1.4 and 4.3 per cent. However, only prior management experience is statistically marginally significant ($\beta = 0.155$, $p < 0.1$). There are no concerns in terms of the model fit because all criteria are in an acceptable range.

No statistically significant relationships could be found between prior entrepreneurial, management, or industry work experience and informal strategy approaches captured by the constructs of flexibility and partnering. Details of the results are displayed the appendix.

Table 7.53: Prior professional experience 1 and futurity

	Model 1	Model 2	Model 3	Model 4
Entrepreneurial experience 1	0.227**			0.149[†]
Management experience 1		0.264***		0.186*
Industry work experience 1			0.150[†]	0.075
R^2	0.051	0.070	0.023	0.094
χ^2	0.3	0.8	1.0	2,3
df	2	2	2	6
P (≥ 0.1)	0.859	0.666	0.620	0.894
χ^2/df (≤ 3.0)	0.152	0.407	0.477	0.376
RMSEA (≤ 0.1)	0.000	0.000	0.000	0.000
CFI (≥ 0.9)	1.000	1.000	1.000	1.000
NFI (≥ 0.9)	0.998	0.996	0.995	0.991
PCLOSE (≥ 0.1)	0.910	0.774	0.730	0.964

[†]$p < 0.1$; *$p < 0.05$; **$p < 0.01$; ***$p < 0.001$

Table 7.54: Prior professional experience 2 and futurity

	Model 1	Model 2	Model 3	Model 4
Entrepreneurial experience 2	0.167*			0.122
Management experience 2		0.207**		0.155^T
Industry work experience 2			0.117	0.057
R^2	0.028	0.043	0.014	0.059
χ^2	1.1	1.6	0.1	2.5
df	2	2	2	6
P ($\geq$ 0.1)	0.577	0.458	0.970	0.870
χ^2/df ($\leq$ 3.0)	0.550	0.780	0.030	0.414
RMSEA ($\leq$ 0.1)	0.000	0.000	0.000	0.000
CFI ($\geq$ 0.9)	1.000	1.000	1.000	1.000
NFI ($\geq$ 0.9)	0.994	0.992	1.000	0.990
PCLOSE ($\geq$ 0.1)	0.706	0.607	0.982	0.954

^{T}p < 0.1; *p < 0.05; **p < 0.01; ***p < 0.001

In addition to the formulated hypotheses, which were based on theory and prior empirical evidence and assumed relationships between prior experience and both formal and informal planning, research question 3.1 asked whether previous professional experience affects other facets of a company's strategic orientation. In this context, the concept of entrepreneurial orientation was considered, which includes the five distinctive dimensions of innovativeness, proactiveness, risk-taking, competitive aggressiveness, and autonomy. However, only the competitive aggressiveness variable was found to be influenced by prior professional experience. Table 7.55 reports statistically significant relationships between prior entrepreneurial (β = 0.219, p < 0.05), management (β = 0.159, p < 0.05), and industry work experience (β = 0.207, p < 0.01), and competitive aggressiveness. These effects could only be found for the first approach, which measured team-level experience with indicator variables. The effect sizes are medium for entrepreneurial and industry work experience and small for management experience. Single variables explain between 2.5 and 4.8 per cent of the variance of competitive aggressiveness. In the full model, 8.2 per cent of the total variance is explained by the explanatory variables, although only industry work experience remains statistically significant (β = 0.196, p < 0.05), with entrepreneurial experience becoming marginally significant (β = 0.141, p < 0.1). Both have small effect sizes. The fit of the full model is acceptable because all the criteria fulfilled the recommended thresholds. The models for the single effects are exactly identified. Therefore, it is not possible to calculate any goodness-of-fit indices.

The detailed analyses of the relationships between prior professional experience and the four remaining dimensions of the entrepreneurial orientation concept are shown in the appendix.

Table 7.55: Prior professional experience 1 and competitive aggressiveness

	Model 1	Model 2	Model 3	Model 4
Entrepreneurial experience 1	0.219*			0.141†
Management experience 1		0.159*		0.062
Industry work experience 1			0.207**	0.196*
R^2	0.048	0.025	0.043	0.082
χ^2				1.0
df	0	0	0	2
P ($\geq$ 0.1)				0.608
χ^2/df ($\leq$ 3.0)				0.497
RMSEA ($\leq$ 0.1)				0.000
CFI ($\geq$ 0.9)				1.000
NFI ($\geq$ 0.9)				0.991
PCLOSE ($\geq$ 0.1)				0.730

†p < 0.1; *p < 0.05; **p < 0.01; ***p < 0.001

Table 7.56: Prior professional experience 2 and competitive aggressiveness

	Model 1	Model 2	Model 3	Model 4
Entrepreneurial experience 2	0.034			0.042
Management experience 2		0.120		0.042
Industry work experience 2			0.100	0.094
R^2	0.001	0.014	0.010	0.016
χ^2				0.8
df	0	0	0	2
P (≥ 0.1)				0.677
χ^2/df (≤ 3.0)				0.390
RMSEA (≤ 0.1)				0.000
CFI (≥ 0.9)				1.000
NFI (≥ 0.9)				0.992
PCLOSE (≥ 0.1)				0.782

$^\dagger p < 0.1$; $^* p < 0.05$; $^{**} p < 0.01$; $^{***} p < 0.001$

7.3.2 Educational background and strategic decision making

In the following, the influence of the educational background of the management team on strategic decision making is examined. Research question 3.2.1 addressed the potential links between educational background and an emphasis on formal planning. Research question 3.2.2 dealt with the impact of education on informal strategy approaches. Finally, research question 3.2.3 concerned the relationship between education and the entrepreneurial strategic posture of the firm. The results revealed no statistically significant relationships between educational background and formal planning activities. More details are given in the appendix. Nevertheless, informal planning methods, reflected by the flexibility and partnering variables, were found to be influenced by executives' education. Tables 7.57 and 7.58 show the corresponding results. Technology education was found to have a statistically significant and positive influence on flexibility ($\beta = 0.187$, $p < 0.05$) and partnering ($\beta = 0.187$, $p < 0.05$), with explained variance of 3.5 per cent in both cases, whereas management education influences partnering only marginally and in a negative direction ($\beta = -0.164$, $p < 0.1$) and explains 3.5 per cent of the variance. All identified effect sizes are small. The model fit could only be estimated for the full models, and all reported criteria are in an acceptable range. The impact of technology education on flexibility remains statistically significant in the full model (β = 0.220, p < 0.05) with a medium effect size. The full model explains 4.9 per cent of the variance.

No statistically significant relationships could be found between the educational background of the management team and the entrepreneurial orientation dimensions of innovativeness, proactiveness, risk-taking, and competitive aggressiveness. Details are given in the appendix.

Table 7.57: Educational background and flexibility

	Model 1	Model 2	Model 3	Model 4
Management education	0.014			0.149[†]
Technology education		0.187*		0.220*
Level of education			0.114	0.086
R^2	0.000	0.035	0.013	0.049
χ^2				0.9
df	0	0	0	2
P ($\geq$ 0.1)				0.630
χ^2/df ($\leq$ 3.0)				0.462
RMSEA ($\leq$ 0.1)				0.000
CFI ($\geq$ 0.9)				1.000
NFI ($\geq$ 0.9)				0.994
PCLOSE ($\geq$ 0.1)				0.747

[†]p < 0.1; *p < 0.05; **p < 0.01; ***p < 0.001

Table 7.58: Educational background and partnering

	Model 1	Model 2	Model 3	Model 4
Management education	-0.164[†]			-0.129[†]
Technology education		0.187*		0.029
Level of education			0.086	0.076
R^2	0.027	0.035	0.007	0.033
χ^2				3.6
df	0	0	0	2
P ($\geq$ 0.1)				0.169
χ^2/df ($\leq$ 3.0)				1.779
RMSEA ($\leq$ 0.1)				0.064
CFI ($\geq$ 0.9)				0.992
NFI ($\geq$ 0.9)				0.984
PCLOSE ($\geq$ 0.1)				0.308

[†]p < 0.1; *p < 0.05; **p < 0.01; ***p < 0.001

7.3.3 Heterogeneity and strategic decision making

The following section examines aspects of the heterogeneity of the executive team such as diversity of composition, educational background, and professional experience. Hypothesis 3.2.2 posits negative impacts of team heterogeneity on the intensity of strategic planning activities. However, the outcomes reveal a marginally significant effect in the opposite direction (β = 0.139, p < 0.1), explaining 1.9 per cent of the variance in the single-factor model. No statistically significant results are obtained in terms of diversity of educational background. Finally, the outcomes regarding the relationship between the diversity of professional experience and planning show a statistically significant positive impact (β = 0.233, p < 0.01), with a medium effect size and an explained variance of 5.4 per cent. Only the latter effect remains statistically significant in the full model (β = 0.215, p < 0.05), with an explained variance of 5.6 per cent. There are no problematic issues in terms of the model fit. Table 7.59 shows the results.

Table 7.59: Heterogeneity and planning

	Model 1	Model 2	Model 3	Model 4
Diversity experience	0.233**			0.215*
Diversity education		0.105		-0.004
Diversity composition			0.139†	0.046
R^2	0.054	0.011	0.019	0.056
χ^2	4.2	2.1	6.3	10.9
df	5	5	5	11
P (≥ 0.1)	0.523	0.841	0.281	0.452
χ^2/df (≤ 3.0)	0.837	0.412	1.255	0.991
RMSEA (≤ 0.1)	0.000	0.000	0.036	0.000
CFI (≥ 0.9)	1.000	1.000	0.993	1.000
NFI (≥ 0.9)	0.979	0.989	0.967	0.962
PCLOSE (≥ 0.1)	0.744	0.934	0.529	0.785

†p < 0.1; *p < 0.05; **p < 0.01; ***p < 0.001

No statistically significant relationships could be found regarding the constructs of futurity, flexibility, and partnering. Details are shown in the appendix. Research question 3.2.3 dealt with the relationship between team diversity and entrepreneurial orientation. All of the examined aspects are statistically significant and positively correlated with competitive aggressiveness. The effect sizes for the explanatory variables of composition ($\beta = 0.242$, p < 0.05), education ($\beta = 0.286$, p < 0.001), and experience ($\beta = 0.272$, p < 0.001) are all medium and explain between 5.8 and 8.2 per cent of the variance of the dependent variable competitive aggressiveness. In the full model, diversity of educational background and prior professional experience remain statistically significant, but with only small effect sizes. In sum, the variables explain 11.7 per cent of the total variance. The fit of the full model is good. Details are shown in table 7.60. Regarding the other dimension of the entrepreneurial orientation concept, only some marginally significant relationships between education, experience, and proactiveness and between experience and risk-taking could be identified. All the effect sizes are small and explain between 2.2 and 2.6 per cent of the variance. An overview is given in the appendix.

Table 7.60: Heterogeneity and competitive aggressiveness

	Model 1	Model 2	Model 3	Model 4
Diversity experience	0.272***			0.187*
Diversity education		0.286***		0.185*
Diversity composition			0.242*	0.052
R^2	0.074	0.082	0.058	0.117
χ^2				1.8
df	0	0	0	2
P (≥ 0.1)				0.403
χ^2/df (≤ 3.0)				0.909
RMSEA (≤ 0.1)				0.000
CFI (≥ 0.9)				1.000
NFI (≥ 0.9)				0.989
PCLOSE (≥ 0.1)				0.557

$^{\dagger}p < 0.1$; $^{*}p < 0.05$; $^{**}p < 0.01$; $^{***}p < 0.001$

7.3.4 Composition and strategic decision making

The composition of the management team was expected to influence the strategic decision making process. As outlined in the second chapter, the team of a research-based spin-off company is very likely to change, and external managers who did not belong to the academic staff of the parent institution often join or replace academic entrepreneurs. Research questions 3.4.1-3.4.3 addressed the potential links between the composition of the management team and formal or informal strategic planning and entrepreneurial orientation. Composition refers to aspects such as external managers on the executive team, ownership of the company, and membership on the original founding team. The results show a marginally significant positive relationship between the proportion of external managers and the intensity of planning activities ($\beta = 0.155$, $p < 0.1$), with a small effect size that accounts for 2.4 per cent of the variance in the model. As reported in table 7.61, there is a statistically marginal negative link between ownership and futurity ($\beta = -0.143$, $p < 0.1$) with a small effect size, explaining 2.0 per cent of the variance. Finally, the proportion of managers who belonged to the original founding team has statistically significant negative impacts on planning ($\beta = -0.249$, $p < 0.01$)

and futurity (β = -0.236, p < 0.01) and explains 6.2 and 5.6 per cent of the variance, respectively. In the full models, only the relationships with the proportion of executives belonging to the founding team remain statistically significant for planning (β = -0.251, p < 0.05) and futurity (β = -0.301, p < 0.01). There are no concerns with goodness-of-fit indices.

Table 7.61: Composition and planning

	Model 1	Model 2	Model 3	Model 4
External manager	0.155[†]			0.001
Ownership		-0.070		0.010
Founder			-0.249**	-0.251*
R^2	0.024	0.005	0.062	0.062
χ^2	2.3	8.9	3.2	9.8
df	5	5	5	11
P ($\geq$ 0.1)	0.805	0.114	0.677	0.547
χ^2/df ($\leq$ 3.0)	0.461	1.777	0.630	0.892
RMSEA ($\leq$ 0.1)	0.000	0.064	0.000	0.000
CFI ($\geq$ 0.9)	1.000	0.978	1.000	1.000
NFI ($\geq$ 0.9)	0.988	0.954	0.984	0.970
PCLOSE ($\geq$ 0.1)	0.917	0.308	0.846	0.844

[†]p < 0.1; *p < 0.05; **p < 0.01; ***p < 0.001

Table 7.62: Composition and futurity

	Model 1	Model 2	Model 3	Model 4
External manager	0.060			-0.154
Ownership		-0.143†		-0.095
Founder			-0.236**	-0.301**
R^2	0.004	0.020	0.056	0.074
χ^2	1.5	1.5	2.7	6.1
df	2	2	2	6
P (≥ 0.1)	0.480	0.475	0.257	0.408
χ^2/df (≤ 3.0)	0.733	0.744	1.360	1.023
RMSEA (≤ 0.1)	0.000	0.000	0.043	0.011
CFI (≥ 0.9)	1.000	1.000	0.996	1.000
NFI (≥ 0.9)	0.992	0.992	0.987	0.981
PCLOSE (≥ 0.1)	0.626	0.621	0.411	0.673

†p < 0.1; *p < 0.05; **p < 0.01; ***p < 0.001

Almost no statistically significant relationships could be identified regarding the impact of team composition on the informal planning aspects of flexibility and partnering. Neither the percentage of external managers within the team nor the ownership structure is related to flexibility or partnering. However, the results indicate a statistically significant negative relationship between the percentage of original founders on the executive team and the partnering construct (β = -0.171, p < 0.05). The effect size is small and responsible for 2.9 per cent of the variance. This relationship remains significant, even with a medium effect size, in the full model (β = -0.212, p < 0.01), which explains 3.1 per cent of the total variance. More details are presented in tables 7.63 and 7.64. The goodness-of-fit indices for the full model demonstrate an appropriate fit, but the single-factor models could not be calculated because they are fully identified.

Table 7.63: Composition and flexibility

	Model 1	Model 2	Model 3	Model 4
External manager	-0.062			-0.160
Ownership		-0.101		-0.102
Founder			-0.022	-0.101
R^2	0.004	0.010	0.001	0.022
χ^2				0.5
df	0	0	0	2
P ($\geq$ 0.1)				0.764
χ^2/df ($\leq$ 3.0)				0.269
RMSEA ($\leq$ 0.1)				0.000
CFI ($\geq$ 0.9)				1.000
NFI ($\geq$ 0.9)				0.997
PCLOSE ($\geq$ 0.1)				0.845

†p < 0.1; *p < 0.05; **p < 0.01; ***p < 0.001

Table 7.64: Composition and partnering

	Model 1	Model 2	Model 3	Model 4
External manager	0.053			-0.100
Ownership		0.024		0.102
Founder			-0.171*	-0.212**
R^2	0.003	0.001	0.029	0.031
χ^2				1.8
df	0	0	0	2
P ($\geq$ 0.1)				0.399
χ^2/df ($\leq$ 3.0)				0.919
RMSEA ($\leq$ 0.1)				0.000
CFI ($\geq$ 0.9)				1.000
NFI ($\geq$ 0.9)				0.993
PCLOSE ($\geq$ 0.1)				0.553

†p < 0.1; *p < 0.05; **p < 0.01; ***p < 0.001

Research question 3.4.3 pointed to a possible relationship between the composition of the management team and the dimensions of the entrepreneurial orientation concept. According to the results displayed in table 7.65, only competitive aggressiveness seems to be influenced by team composition. The outcomes of the analysis indicate a statistically significant and medium-sized negative effect of founders within the team on the level of competitive aggressiveness (β = -0.241, p < 0.01). The explained variance is 5.8 per cent. A marginally significant relationship with the proportion of external managers can be observed (β = 0.198, p < 0.1), with a small positive effect size and an explained variance of 3.9 per cent. In the full model, the variable capturing the percentage of founders is only marginally significant, with a medium effect size (β = -0.203, p < 0.1) and an explained variance of 6.2 per cent. This result might be caused by the negative correlation between the percentage of external managers and the proportion of team members who belonged to the original founding team (p < 0.01). The overall model fit is good.

Table 7.65: Composition and competitive aggressiveness

	Model 1	Model 2	Model 3	Model 4
External manager	0.198†			0.071
Ownership		-0.064		0.011
Founder			-0.241**	-0.203†
R^2	0.039	0.004	0.058	0.062
χ^2				1.0
df	0	0	0	2
P ($\geq$ 0.1)				0.621
χ^2/df ($\leq$ 3.0)				0.477
RMSEA ($\leq$ 0.1)				0.000
CFI ($\geq$ 0.9)				1.000
NFI ($\geq$ 0.9)				0.995
PCLOSE ($\geq$ 0.1)				0.740

†p < 0.1; *p < 0.05; **p < 0.01; ***p < 0.001

7.3.5 General attributes and strategic decision making

The next section presents the results regarding the links between demographic attributes and general characteristics of the executive team and strategic decision processes, which were addressed by research questions 3.5.1-3.5.3. The analysis considers the size, average age, and gender distribution of the management team. The results, displayed in tables 7.66 and 7.67, show statistically significant relationships between team size and formal planning, including both strategic planning activities and long-term firm orientation as captured by the futurity construct. Team size has a positive effect on the level of strategic planning activities (β = 0.200, p < 0.05), with a medium effect size, explaining 4.0 per cent of the variance. This finding remains statistically significant in the full model (β = 0.202, p < 0.05). All reports yielded appropriate fit indices.

Table 7.66: General attributes and planning

	Model 1	Model 2	Model 3	Model 4
Team size	0.200**			0.202**
Average age		0.037		0.029
Gender distribution			0.086	0.090
R^2	0.040	0.001	0.007	0.049
χ^2	2.5	5.1	8.7	12.1
df	5	5	5	11
P ($\geq$ 0.1)	0.779	0.404	0.124	0.359
χ^2/df ($\leq$ 3.0)	0.497	1.020	1.731	1.097
RMSEA ($\leq$ 0.1)	0.000	0.010	0.062	0.022
CFI ($\geq$ 0.9)	1.000	0.999	0.979	0.994
NFI ($\geq$ 0.9)	0.987	0.973	0.955	0.941
PCLOSE ($\geq$ 0.1)	0.903	0.648	0.324	0.716

[†]p < 0.1; *p < 0.05; **p < 0.01; ***p < 0.001

The size of the management team also positively influences long-term planning, as indicated by a medium effect size (β = 0.264, p < 0.001) and explained variance of 7.0 per cent. The effect remains statistically significant in the full model (β = 0.266, p < 0.001), where it explains 9.1 per cent of the variance in the dependent variable futurity. The fit indices for the models including the size of the management team correspond to the discussed recommendations. All details are shown in table 7.67.

Table 7.67: General attributes and futurity

	Model 1	Model 2	Model 3	Model 4
Team size	0.264***			0.266***
Average age		-0.034		-0.034
Gender distribution			0.083	0.091
R^2	0.070	0.001	0.007	0.078
χ^2	1.0	6.2	0.7	7.8
df	2	2	2	6
P ($\geq$ 0.1)	0.608	0.045	0.718	0.256
χ^2/df ($\leq$ 3.0)	0.498	3.112	0.332	1.294
RMSEA ($\leq$ 0.1)	0.000	0.105	0.000	0.039
CFI ($\geq$ 0.9)	1.000	0.978	1.000	0.991
NFI ($\geq$ 0.9)	0.995	0.969	0.997	0.964
PCLOSE ($\geq$ 0.1)	0.730	0.119	0.812	0.525

†p < 0.1; *p < 0.05; **p < 0.01; ***p < 0.001

Only the competitive aggressiveness dimension of the entrepreneurial orientation concept was found to be positively influenced by the size of the management team. The results demonstrate a statistically significant relationship with a small effect size (β = 0.167, p < 0.01) and an explained variance of 2.8 per cent. Team size is also statistically significant in the full model, with a medium effect size (β = 0.214, p < 0.01). The fit of the full model is in an adequate range, whereas the single models are fully identified, so fit indices cannot be estimated. The results are presented in table 7.68. Furthermore, the average age of the management team members is found to be statistically significantly related to flexibility (β = -0.277, p < 0.01), and the proportion of males in the team has a statistically significant impact on the entrepreneurial orientation dimension of proactiveness (β = 0.190, p < 0.05). However, the skewness and kurtosis of both influence factors were in the critical range. All details are displayed in the appendix.

Table 7.68: General attributes and competitive aggressiveness

	Model 1	Model 2	Model 3	Model 4
Team size	0.167**			0.214**
Average age		-0.080		-0.085
Gender distribution			0.161	0.122†
R^2	0.028	0.006	0.026	0.064
χ^2				1.3
df	0	0	0	2
P ($\geq$ 0.1)				0.535
χ^2/df ($\leq$ 3.0)				0.626
RMSEA ($\leq$ 0.1)				0.000
CFI ($\geq$ 0.9)				1.000
NFI ($\geq$ 0.9)				0.982
PCLOSE ($\geq$ 0.1)				0.672

$^{\dagger}p < 0.1$; $^*p < 0.05$; $^{**}p < 0.01$; $^{***}p < 0.001$

7.3.6 Overview of executive characteristics and strategic decision making

The previous sections provided an overview of the findings concerning the impact of team characteristics on strategic decision making. Some notable observations are summarised in table 7.69. Prior professional experience was found to influence formal planning activities positively, but no links could be identified between experience and informal strategy approaches. In addition, heterogeneity of former experience is positively associated with the level of strategic planning activities. A high percentage of team members educated in a technology-oriented field is positively related to informal strategy approaches, reflected by the flexibility and partnering constructs. The proportion of team members who belonged to the original founding team negatively affects both aspects of formal strategy making and the partnering latent variable. The size of the management team is positively linked to formal strategising as well. Interestingly, only competitive aggressiveness, as a dimension of the entrepreneurial orientation of the firm, was influenced by several factors, such as prior professional experience, heterogeneity, proportion of founders, and team size. There are almost no significant relationships between team characteristics and the other entrepreneurial orientation dimensions of innovativeness, proactiveness, and risk-taking.

Table 7.69: Overview of executive characteristics and strategic decision making

		PLAN	FUT	FLEX	PART	INNO	PRO	RISK	AGG
Prior professional experience	Entrepreneurial experience 1	+**	+**						+*
	Management experience 1	+**	+***						+*
	Industry work experience 1	+**	+†						+**
Educational background	Management education					-†			
	Technology education			+*	+*				
	Level of education								
Heterogeneity	Diversity experience	+**					+†	+†	+***
	Diversity education						+†		+***
	Diversity composition	+†							+*
Composition	External manager	+†							+†
	Ownership		-†						
	Founder	-**	-**			-*			-**
Demographic attributes and general characteristics	Team size	+**	+***						+**
	Average age		-**						
	Gender distribution						+*	+†	

†p < 0.1; *p < 0.05; **p < 0.01; ***p < 0.001

7.4 Link between firm and environmental factors and strategic decision making

In addition to the characteristics of the management team that were discussed in the previous sections, the impact of several firm-specific and environmental characteristics on strategic decision making on the organisational level are analysed in the next step. The maturity of the ventures is captured by aspects such as the age, life-cycle stage, and size of the company. The technology orientation, research intensity, and product focus variables reflect how strongly a firm's portfolio is based on high technology and intensive research efforts and whether the firm is service or product oriented. The level of environmental uncertainty was measured by the state and effect uncertainty variables, which distinguish between potential changes to the general environment and possible effects on the companies.

7.4.1 Company maturity and strategic decision making

A descriptive overview of the firm age, life-cycle stage, and firm size variables, which are related to the maturity of the companies, is given in table 7.70. On average, the ventures are around 9.5 years old and have almost 20 employees. As the Kolmogorov-Smirnov test shows, the assumption of a normal distribution seems to be unproblematic for the firm age variable because the significance test does not reject the null hypothesis ($p < 0.05$). On the other hand, the distribution of the firm size variable, measured by the number of employees, shows a high absolute value for kurtosis of 13.663, which is above the recommended threshold (Kline, 2005). The skewness and kurtosis of firm age and life-cycle stage are in an acceptable range. Thus, the results for the firm size variable are only briefly discussed in the following section because any conclusions might be questionable.

Table 7.70: Descriptive statistics of company maturity

	N	Mean	SD	KS	P (KS)	Skewness	Kurtosis
Firm age	189	9.526	4.656	0.985	0.286	0.607	0.172
Life-cycle stage	190	2.34	0.944	4.170	0.000	0.652	-0.031
Firm size	154	19.65	25.360	2.913	0.000	3.207	13.663

SD: Standard deviation, KS: Kolmogorov-Smirnov-test

Hypotheses 4.1.1 and 4.1.2 stated that companies in a later phase of their development emphasise formal and deemphasise informal planning activities. In this respect, the results are contradictory. Firm age has a statistically significant positive impact on flexibility (β = 0.118, $p < 0.05$), with a small effect size, explaining 1.4 per cent of the variance. This effect remains statistically significant in the full model. None of the other factors has a significant influence. The size of the firm is statistically significantly and positively associated with futurity (β = 0.310, $p < 0.001$), representing long-term planning and analysing, and explains 9.6 per cent of the variance. The effect remains significant in the full model (β = 0.310, $p < 0.001$), with an explained variance of 11.1 per cent. None of the other variables is statistically significantly related to futurity. However, firm size is also positively related to partnering (β = 0.163, $p < 0.05$), with a small effect size responsible for 2.6 per cent of the variance. This effect remains statistically significant in the full model as well (β = 0.241, $p < 0.05$). Partnering was described as an informal strategy approach emphasising partnerships and collaborations. Table 7.71 shows the results for the relationship between company maturity and flexibility. As mentioned above, the details for the other effects are displayed in the appendix.

Table 7.71: Company maturity and flexibility

	Model 1	Model 2	Model 3	Model 4
Firm age	0.118*			0.113*
Life-cycle stage		0.024		-0.064
Firm size (employees)			0.066	-0.010
R^2	0.014	0.001	0.004	0.010
χ^2				1.6
df	0	0	0	2
P ($\geq$ 0.1)				0.460
χ^2/df ($\leq$ 3.0)				0.777
RMSEA ($\leq$ 0.1)				0.000
CFI ($\geq$ 0.9)				1.000
NFI ($\geq$ 0.9)				0.990
PCLOSE ($\geq$ 0.1)				0.608

$^{\dagger}p < 0.1$; $^{*}p < 0.05$; $^{**}p < 0.01$; $^{***}p < 0.001$

Research question 4.1 asked how the maturity of the company influences a firm's strategic posture. The results reveal different effects across the entrepreneurial orientation dimensions. Firm size has a statistically significant positive impact on competitive aggressiveness (β = 0.354, p < 0.001), with a medium effect size and an explained variance of 12.5 per cent. The full model also shows a significant link (β = 0.441, p < 0.001), with a slightly larger effect size. The life-cycle stage is statistically significantly and negatively related to proactiveness (β = -0.193, p < 0.05) and risk-taking (β = -0.253, p < 0.01), explaining variance of 3.7 and 6.4 per cent, respectively. Both effects remain statistically significant in the full models, with increased effect sizes for proactiveness (β = -0.307, p < 0.01) and risk-taking (β = -0.326, p < 0.01). The results for the risk-taking dependent variable are shown in table 7.72. Further details are given in the appendix.

Table 7.72: Company maturity and risk-taking

	Model 1	Model 2	Model 3	Model 4
Firm age	-0.028			0.098
Life-cycle stage		-0.253**		-0.326**
Firm size (employees)			-0.043	0.056
R^2	0.001	0.064	0.002	0.080
χ^2	2.3	0.4	5.0	8.4
df	2	2	2	6
P ($\geq$ 0.1)	0.324	0.801	0.080	0.212
χ^2/df ($\leq$ 3.0)	1.128	0.222	2.523	1.395
RMSEA ($\leq$ 0.1)	0.026	0.000	0.089	0.045
CFI ($\geq$ 0.9)	0.995	1.000	0.942	0.981
NFI ($\geq$ 0.9)	0.962	0.993	0.920	0.942
PCLOSE ($\geq$ 0.1)	0.481	0.871	0.182	0.473

[†]p < 0.1; *p < 0.05; **p < 0.01; ***p < 0.001

7.4.2   Environmental uncertainty and strategic decision making

Two different aspects of environmental uncertainty were taken into consideration. State uncertainty reflects whether certain aspects of the environmental situation can be predicted, whereas effect uncertainty refers to the potential effect on the company of environmental changes (McKelvie, Haynie, and Gustavsson, 2011). Table 7.73 gives a descriptive overview of the perceived environmental uncertainty for the respondents in terms of state and effect uncertainty. On average, respondents perceived more state uncertainty than effect uncertainty. According to the Kolmogorov-Smirnov-test, the distributions do not perfectly follow a normal distribution. Nevertheless, the maximum absolute values of 0.399 for skewness and 0.223 for kurtosis are in an acceptable range and violate the assumption of a normal distribution only moderately.

Table 7.73: Descriptive statistics of environmental uncertainty

	N	Mean	SD	KS	P (KS)	Skewness	Kurtosis
State uncertainty	185	4.468	1.075	1.384	0.043	-0.399	0.016
Effect uncertainty	185	3.978	1.155	1.847	0.002	0.131	-0.223

SD: Standard deviation, KS: Kolmogorov-Smirnov-test

According to hypotheses 4.2.2 and 4.2.3, uncertainty should decrease the intensity of formal planning and increase informal planning activities. The impacts of both dimensions of environmental uncertainty on the dependent variables planning and futurity are displayed in tables 7.74 and 7.75. The results show a statistically significant negative relationship between effect uncertainty and the level of strategic planning (β = -0.314, p < 0.001), with a medium effect size, explaining 9.9 per cent of the variance. The effect uncertainty variable also negatively influences the long-term firm orientation, captured by the construct futurity, with a medium effect size (β = -0.264, p < 0.001). The factor is responsible for 7.0 per cent of the variance of the dependent variable. Both effects remain statistically significant in the full model, where all factors together explain 10.0 and 10.1 per cent of the variance, respectively. However, no statistically significant relationship could be found between state uncertainty and planning. The goodness-of-fit indices are in the proper range for almost all models.

Table 7.74: Environmental uncertainty and planning

	Model 1	Model 2	Model 3
State uncertainty	-0.098		0.034
Effect uncertainty		-0.314***	-0.327***
R^2	0.010	0.099	0.100
χ^2	8.8	7.0	13.5
df	5	5	8
P ($\geq$ 0.1)	0.119	0.222	0.096
χ^2/df ($\leq$ 3.0)	1.753	1.397	1.685
RMSEA ($\leq$ 0.1)	0.063	0.045	0.060
CFI ($\geq$ 0.9)	0.979	0.990	0.975
NFI ($\geq$ 0.9)	0.955	0.966	0.944
PCLOSE ($\geq$ 0.1)	0.316	0.461	0.334

†p < 0.1; *p < 0.05; **p < 0.01; ***p < 0.001

Table 7.75: Environmental uncertainty and futurity

	Model 1	Model 2	Model 3
State uncertainty	0.053		0.194*
Effect uncertainty		-0.264***	-0.340***
R^2	0.003	0.070	0.101
χ^2	4.9	1.7	6.8
df	2	2	4
P ($\geq$ 0.1)	0.087	0.432	0.145
χ^2/df ($\leq$ 3.0)	2.446	0.840	1.710
RMSEA ($\leq$ 0.1)	0.087	0.000	0.061
CFI ($\geq$ 0.9)	0.985	0.992	0.988
NFI ($\geq$ 0.9)	0.975	1.000	0.972
PCLOSE ($\geq$ 0.1)	0.192	0.583	0.333

†p < 0.1; *p < 0.05; **p < 0.01; ***p < 0.001

The effect of perceived environmental uncertainty on informal strategy aspects is displayed in tables 7.76 and 7.77. According to the presented results, state uncertainty has a statistically significant positive impact on flexibility with a medium effect size (β = 0.235, p < 0.01) and an explained variance of 5.5 per cent. The effect becomes only marginally significant in the full model (β = 0.230, p < 0.1). There is no statistically significant relationship between effect uncertainty and flexibility. On the other hand, effect uncertainty negatively influences the partnering variable but is only marginally significant (β = -0.191, p < 0.1). There seems to be no link between state uncertainty and partnering. However, the fit indices for the full models do not meet the recommendations.

Table 7.76: Environmental uncertainty and flexibility

	Model 1	Model 2	Model 3
State uncertainty	0.235**		0.230†
Effect uncertainty		0.028	-0.124
R^2	0.055	0.001	0.046
χ^2			6.3
df	0	0	1
P ($\geq$ 0.1)			0.012
χ^2/df ($\leq$ 3.0)			6.307
RMSEA ($\leq$ 0.1)			0.166
CFI ($\geq$ 0.9)			0.955
NFI ($\geq$ 0.9)			0.950
PCLOSE ($\geq$ 0.1)			0.035

†p < 0.1; *p < 0.05; **p < 0.01; ***p < 0.001

Table 7.77: Environmental uncertainty and partnering

	Model 1	Model 2	Model 3
State uncertainty	-0.050		-0.057
Effect uncertainty		-0.191†	-0.142
R^2	0.003	0.037	0.030
χ^2			3.6
df	0	0	1
P (≥ 0.1)			0.058
χ^2/df (≤ 3.0)			3.589
RMSEA (≤ 0.1)			0.116
CFI (≥ 0.9)			0.985
NFI (≥ 0.9)			0.980
PCLOSE (≥ 0.1)			0.120

†p < 0.1; *p < 0.05; **p < 0.01; ***p < 0.001

Research question 4.2 addressed the relationships between perceived environmental uncertainty and the firm's entrepreneurial strategic posture. Tables 7.78 and 7.79 report the results for the innovativeness and proactiveness dimensions. State uncertainty was found to have a statistically significant positive influence on innovativeness ($\beta = 0.217$, p < 0.05), with a medium effect size, explaining 4.7 per cent of the variance. This effect remains statistically significant in the full model ($\beta = 0.287$, p < 0.05). Effect uncertainty has no statistically significant effect as a single factor, but it becomes marginally significant in the full model ($\beta = -0.178$, p < 0.1), with a negative impact and a small effect size. The full model explains 7.4 per cent of the total variance and is not problematic in terms of the reported fit indices. Effect uncertainty has a statistically significant negative impact on the entrepreneurial orientation dimension of proactiveness ($\beta = -0.214$, p < 0.05). The effect size is medium and the explained variance is 4.6 per cent. This relationship remains statistically significant in the full model ($\beta = -0.267$, p < 0.01), which explains 6.1 per cent of the overall variance. All of the goodness-of-fit indices are in line with the discussed recommendations. No statistically significant effects could be found for the links between perceived environmental uncertainty and the risk-taking and competitive aggressiveness dimensions.

Table 7.78: Environmental uncertainty and innovativeness

	Model 1	Model 2	Model 3
State uncertainty	0.217*		0.287*
Effect uncertainty		-0.070	-0.178†
R^2	0.047	0.005	0.074
χ^2			0.022
df	0	0	1
P ($\geq$ 0.1)			0.882
χ^2/df ($\leq$ 3.0)			0.022
RMSEA ($\leq$ 0.1)			0.000
CFI ($\geq$ 0.9)			1.000
NFI ($\geq$ 0.9)			1.000
PCLOSE ($\geq$ 0.1)			0.907

†p < 0.1; *p < 0.05; **p < 0.01; ***p < 0.001

Table 7.79: Environmental uncertainty and proactiveness

	Model 1	Model 2	Model 3
State uncertainty	0.039		0.135
Effect uncertainty		-0.214*	-0.267**
R^2	0.002	0.046	0.061
χ^2	2.5	2.3	5.0
df	2	2	4
P ($\geq$ 0.1)	0.281	0.324	0.286
χ^2/df ($\leq$ 3.0)	1.270	1.127	1.252
RMSEA ($\leq$ 0.1)	0.038	0.026	0.036
CFI ($\geq$ 0.9)	0.994	0.998	0.992
NFI ($\geq$ 0.9)	0.976	0.980	0.966
PCLOSE ($\geq$ 0.1)	0.437	0.482	0.509

†p < 0.1; *p < 0.05; **p < 0.01; ***p < 0.001

7.4.3   Technology base and strategic decision making

Because the levels of technology and research intensity play an important role in the discussion of new technology-oriented ventures and academic spin-offs, research questions 4.3.1-4.3.3 refer to the link between these aspects and strategic decision making. Descriptive statistics are displayed in table 7.80. On average, respondents reported high levels of technology orientation and research intensity. Because the respondents work in spin-offs that emerged from research institutes, this finding is not surprising. The standard deviation is smallest for the technology orientation measure and highest for the variable indicating whether the portfolio is based on services or products. The Kolmogorov-Smirnov test does not confirm exact normal distributions of the variables, but the skewness and kurtosis have maximum absolute values of 1.791 and 3.266, respectively, and are still tolerable.

Table 7.80: Descriptive statistics of technology base

	N	Mean	SD	KS	P (KS)	Skewness	Kurtosis
Technology orientation	190	6.18	1.174	4.042	0.000	-1.791	3.266
Research intensity	188	5.47	1.542	3.072	0.000	-0.944	0.090
Product focus	188	4.29	1.886	2.297	0.000	-0.304	-1.062

SD: Standard deviation, KS: Kolmogorov-Smimov-test

The analysis examined the potential impacts on formal and informal planning activities as well as the entrepreneurial strategic posture. The results show a clear pattern. On one hand, there is almost no statistically significant impact on formal or informal planning. Only the technology orientation seems to have a statistically significant positive impact on the partnering variable (β = 0.252, p < 0.05), demonstrating a medium effect size with an explained variance of 1.8 per cent. Details are given in the appendix. On the other hand, there are several statistically significant relationships between technology orientation, research intensity, product focus, and the entrepreneurial orientation dimensions of innovativeness, proactiveness, risk-taking, and competitive aggressiveness.

Table 7.81 presents the results for the dependent variable innovativeness, which is statistically significantly and positively related to technology orientation (β = 0.295, p < 0.001) and research intensity (β = 0.265, p < 0.001). The factors explain 10.6 and 7.0 per cent of the variance, respectively. In the full model, only the impact of technology orientation remains statistically significant (β = 0.292, p < 0.001). The full model explains 12.8 per cent of the variance and has a good model fit.

Table 7.81: Technology base and innovativeness

	Model 1	Model 2	Model 3	Model 4
Technology orientation	0.295***			0.292**
Research intensity		0.265*		0.112
Product focus			0.136	-0.002
R^2	0.106	0.070	0.018	0.128
χ^2				1.1
df	0	0	0	2
P ($\geq$ 0.1)				0.583
χ^2/df ($\leq$ 3.0)				0.539
RMSEA ($\leq$ 0.1)				0.000
CFI ($\geq$ 0.9)				1.000
NFI ($\geq$ 0.9)				0.990
PCLOSE ($\geq$ 0.1)				0.710

†p < 0.1; *p < 0.05; **p < 0.01; ***p < 0.001

According to the figures displayed in table 7.82, all of the explanatory variables are statistically significantly associated with the dimension of proactiveness. Technology orientation has the greatest impact, with a large effect size (β = 0.500, p < 0.001), and is responsible for 25.0 per cent of the variance. Research intensity also positively influences the level of proactiveness in the company (β = 0.171, p < 0.001). The effect size is small, and the explained variance is 12.5 per cent. Finally, the product orientation factor has a positive impact with a medium effect size (β = 0.328, p < 0.001) and an explained variance of 10.8 per cent. In the full model, technology orientation and product focus remain statistically significant. All of the factors together explain 30.0 per cent of the total variance.

Table 7.82: Technology base and proactiveness

	Model 1	Model 2	Model 3	Model 4
Technology orientation	0.500***			0.397***
Research intensity		0.171***		0.115
Product focus			0.328***	0.177*
R^2	0.250	0.125	0.108	0.300
χ^2	1.0	1.4	1.8	3.1
df	2	2	2	6
P ($\geq$ 0.1)	0.503	0.507	0.411	0.791
χ^2/df ($\leq$ 3.0)	0.605	0.678	0.890	0.523
RMSEA ($\leq$ 0.1)	0.000	0.000	0.000	0.000
CFI ($\geq$ 0.9)	1.000	1.000	1.000	1.000
NFI ($\geq$ 0.9)	0.993	0.989	0.985	0.986
PCLOSE ($\geq$ 0.1)	0.727	0.649	0.564	0.919

†p < 0.1; *p < 0.05; **p < 0.01; ***p < 0.001

Table 7.83 clearly shows that all the factors have a statistically significant positive impact on risk-taking with a medium effect size. Technology orientation (β = 0.476, p < 0.001) explains 22.7 per cent of the variance, whereas research intensity (β = 0.474, p < 0.001) and product focus (β = 0.360, p < 0.001) explain 22.4 and 12.9 per cent, respectively, in the single models. All of the explanatory variables remain statistically significant in the full model, explaining a total of 33.7 per cent of the variance. None of the goodness-of-fit indices seem to be problematic, indicating appropriate model fits.

In terms of the entrepreneurial orientation dimension of competitive aggressiveness, all of the independent variables have a positive effect. Technology orientation shows a small effect size (β = 0.090, p < 0.05), whereas research intensity (β = 0.209, p < 0.05) and product focus (β = 0.285, p < 0.001) have medium effect sizes. The factors each explain between 0.8 and 8.1 per cent of the variance of competitive aggressiveness. Only product focus is statistically significant in the full model, with an explained variance of 9.7 per cent. The model fit is in the appropriate range.

Table 7.83: Technology base and risk-taking

	Model 1	Model 2	Model 3	Model 4
Technology orientation	0.476***			0.295**
Research intensity		0.474***		0.275**
Product focus			0.360***	0.185*
R^2	0.227	0.224	0.129	0.337
χ^2	0.5	0.4	0.1	1.1
df	2	2	2	6
P (≥ 0.1)	0.768	0.819	0.954	0.982
χ^2/df (≤ 3.0)	0.264	0.199	0.047	0.182
RMSEA (≤ 0.1)	0.000	0.000	0.000	0.000
CFI (≥ 0.9)	1.000	1.000	1.000	1.000
NFI (≥ 0.9)	0.994	0.995	0.999	0.994
PCLOSE (≥ 0.1)	0.848	0.883	0.971	0.995

†p < 0.1; *p < 0.05; **p < 0.01; ***p < 0.001

Table 7.84: Technology base and competitive aggressiveness

	Model 1	Model 2	Model 3	Model 4
Technology orientation	0.090*			0.054
Research intensity		0.209*		0.099
Product focus			0.285***	0.236**
R^2	0.008	0.044	0.081	0.097
χ^2				2.5
df	0	0	0	2
P (≥ 0.1)				0.284
χ^2/df (≤ 3.0)				1.260
RMSEA (≤ 0.1)				0.037
CFI (≥ 0.9)				0.996
NFI (≥ 0.9)				0.982
PCLOSE (≥ 0.1)				0.440

†p < 0.1; *p < 0.05; **p < 0.01; ***p < 0.001

7.4.4　Overview of firm and environmental factors and strategic decision making

The findings regarding the links between firm characteristics, environmental uncertainty and strategic decision processes are presented in table 7.85. There is no consistent pattern for the relationships between the variables representing the maturity of the company and the analysed strategy approaches. All of the variables referring to the technology base have a significant positive impact on most of the entrepreneurial orientation dimensions, but they have almost no effect on formal or informal strategic decision making. The two distinguished facets of perceived environmental uncertainty are differently associated with the strategy constructs. The effect uncertainty variable, which considers possible consequences for the firm, has a negative impact on the level of strategic planning activities and on the long-term firm orientation, both of which represent formal and rational aspects in the context of strategic decision making. State uncertainty is positively related to flexibility and the entrepreneurial orientation dimension of innovativeness.

Table 7.85: Overview of firm and environmental characteristics and strategic decision making

		PLAN	FUT	FLEX	PART	INNO	PRO	RISK	AGG
Company maturity	Firm age			+*					
	Life-cycle stage					+†	-*	-**	
	Firm size (employees)	+***			+*				+***
Environmental uncertainty	State uncertainty			+**		+*			
	Effect uncertainty	-***	-***		-†		-*		
Technology base	Technology orientation				+**	+***	+***	+***	+*
	Research intensity	+†				+*	+***	+***	+*
	Product focus	+†	+†			+***	+***	+***	

†p < 0.1; *p < 0.05; **p < 0.01; ***p < 0.001

7.5   Integrated models

The previous sections have outlined the relationships between executive team characteristics and performance, between aspects of strategic decision making processes and performance, and between team characteristics, firm and environmental factors, and strategy making. The next question is how certain factors interact. In the following section integrated models including several influence factors are discussed.

7.5.1   Discriminant analysis

The analysis of the first link between team characteristics and performance yielded very few statistically significant relationships. Thus, a broader model that includes several factors does not contribute to this study. On the contrary, examination of the second link confirmed that several strategy process variables have a statistically significant impact on firm performance. These include the latent variables of planning, futurity, innovativeness, proactiveness, and competitive aggressiveness. Before combining these factors in an overall model, the following analysis will determine whether the constructs are capturing distinctive phenomena. For this purpose, discriminant validity can be assessed by applying the chi-squared difference test, which compares the original model with a new model in which the correlation between two latent variables is fixed to 1. In this context, the discriminant validity of the underlying constructs was evaluated by the Fornell-Larcker criterion, which is generally stricter than the chi-squared difference test and requires the AVE of a factor to be higher than the squared correlations of this factor with all other factors in the model (Fornell and Larcker, 1981). As displayed in table 7.86, the planning variable is similar to futurity, and innovativeness is similar to proactiveness. Consequently, planning and futurity have to be discussed in separate models. Innovativeness will not be considered in further analyses because the fit of the measurement model was poor. The latent variable proactiveness will be included in the integrated model instead.

Table 7.86: Discriminant analysis with Fornell-Larcker criterion

	PLAN	FUT	PART	INNO	PRO	AGG
Planning	**0.458**					
Futurity	0.573	**0.579**				
Partnering	0.235	0.246	**0.730**			
Innovativeness	0.000	0.021	0.086	**0.347**		
Proactiveness	0.045	0.044	0.012	0.347	**0.455**	
Competitive aggressiveness	0.080	0.036	0.045	0.068	0.089	**0.567**

Additionally, the analysis of discriminant validity on the indicator level confirms that all indicators demonstrate the highest correlations with the latent variable to which they are supposed to contribute. Table 7.87 displays the detailed results.

Table 7.87: Discriminant validity on the indicator level

	PLAN	FUT	FLEX	PAR	INNO	PRO	RISK	AGG
PLAN 1	**0.708**	0.426	-0.103	0.160	-0.006	0.105	0.171	0.178
PLAN 2	**0.747**	0.378	0.049	0.247	0.066	0.190	0.221	0.264
PLAN 3	**0.799**	0.546	-0.053	0.347	0.063	0.144	0.191	0.267
PLAN 4	**0.804**	0.507	-0.016	0.366	0.017	0.033	0.217	0.192
FUT 2	0.548	**0.876**	-0.045	0.366	0.084	0.141	0.171	0.219
FUT 3	0.519	**0.887**	0.013	0.361	0.141	0.119	0.100	0.106
FUT 4	0.463	**0.742**	0.076	0.396	0.064	0.144	0.121	0.161
FLEX 1	-0.060	-0.047	**0.893**	0.126	0.234	0.157	-0.018	-0.121
FLEX 3	0.004	0.081	**0.893**	0.197	0.123	0.177	0.086	-0.049
PAR 1	0.322	0.424	0.178	**0.926**	0.212	0.095	0.120	0.066
PAR 2	0.366	0.397	0.164	**0.926**	0.237	0.048	0.138	0.129
INNO 2	-0.014	0.086	0.165	0.169	**0.821**	0.375	0.111	0.080
INNO 3	0.110	0.097	0.145	0.217	**0.821**	0.201	0.258	0.191
PRO 1	0.087	0.089	0.140	-0.019	0.229	**0.761**	0.203	0.177
PRO 2	0.162	0.189	0.137	0.091	0.281	**0.812**	0.303	0.169
PRO 3	0.080	0.080	0.148	0.109	0.347	**0.804**	0.310	0.117
RISK 1	0.206	0.122	0.023	0.156	0.133	0.395	**0.771**	0.185
RISK 2	0.119	0.043	0.024	0.002	0.211	0.181	**0.747**	0.223
RISK 3	0.242	0.159	0.011	0.147	0.137	0.177	**0.705**	0.381
AGG 1	0.340	0.245	-0.026	0.215	0.262	0.308	0.396	**0.866**
AGG 2	0.180	0.119	-0.136	-0.002	0.076	0.059	0.239	**0.866**

7.5.2   Strategic decision processes and firm performance

The first model includes the strategic orientation dimensions of planning, partnering, proactiveness, and competitive aggressiveness as explanatory variables. Dependent variables are the performance dimensions of growth and profitability. In the growth model, only planning (β = 0.262, p < 0.05) and proactiveness (β = 0.344, p < 0.001) remain statistically significant, and both have medium effect sizes. In combination, the factors explain 27.1 per cent of the variance. In terms of the model fit, only the probability value of the chi-squared test of goodness-of-fit is not reached. In the model including profitability as the dependent variable, only partnering remains statistically significant (β = 0.266, p < 0.05), with a medium effect size. Proactiveness demonstrates a marginally positive influence (β = 0.179, p < 0.1), with a small effect size. In sum, the factors explain 9.8 per cent of the variance. Details are shown in table 7.88.

Table 7.88: Integrated model of strategic decision processes and performance 1

	GROWTH	PROFIT
Planning	0.262*	-0.020
Partnering	0.092	0.266*
Proactiveness	0.344***	0.179†
Competitive aggressiveness	0.022	-0.048
R^2	0.271	0.098
Model fit:		
χ^2	93.2	75.1
df	67	55
P ($\geq$ 0.1)	0.019	0.037
χ^2/df ($\leq$ 3.0)	1.391	1.365
RMSEA ($\leq$ 0.1)	0.045	0.044
CFI ($\geq$ 0.9)	0.970	0.970
NFI ($\geq$ 0.9)	0.904	0.900
PCLOSE ($\geq$ 0.1)	0.626	0.649

†p < 0.1; *p < 0.05; **p < 0.01; ***p < 0.001

The second model replaces the planning construct with the futurity latent variable. As displayed in table 7.89, this analysis yields similar results. Futurity (β = 0.219, p < 0.05) and proactiveness (β = 0.347, p < 0.001) each have a statistically significant positive impact on the performance dimension of growth. Both effect sizes are medium. All of the factors in the model combined explain 24.9 per cent of the total variance. The overall model fit seems to be acceptable. Only the probability value of the chi-squared test of goodness-of-fit is below the recommended threshold. The model using profitability as the dependent variable reports similar results to those outlined in the previous paragraph. The partnering variable has a statistically significant positive impact on profitability, with a medium effect size (β = 0.267, p < 0.05), whereas the proactiveness construct is marginally significantly related and has a small effect size (β = 0.182, p < 0.1). In combination, the independent variables explain 9.9 per cent of the total variance. Except for the probability value of the chi-squared test, all of the goodness-of-fit indices confirm a proper model fit.

Table 7.89: Integrated model of strategic decision processes and performance 2

	GROWTH	PROFIT
Futurity	0.219*	-0.043
Partnering	0.096	0.267*
Proactiveness	0.347***	0.182†
Competitive aggressiveness	0.037	-0.038
R^2	0.249	0.099
Model fit:		
χ^2	94.5	75.7
df	55	44
P ($\geq$ 0.1)	0.001	0.002
χ^2/df ($\leq$ 3.0)	1.718	1.719
RMSEA ($\leq$ 0.1)	0.061	0.061
CFI ($\geq$ 0.9)	0.955	0.953
NFI ($\geq$ 0.9)	0.903	0.900
PCLOSE ($\geq$ 0.1)	0.183	0.204

†p < 0.1; *p < 0.05; **p < 0.01; ***p < 0.001

7.5.3   Antecedents and consequences of strategic decision processes

In the next step, predictors of strategic orientation dimensions are included in the model. As shown in the previous paragraphs, the impacts of planning, futurity, and proactiveness on growth remained statistically significant in the full models. Thus, the planning, futurity, and proactiveness variables will be considered for their impact on growth, but planning and futurity will be analysed separately. Those predictors on the team, firm, and environmental levels will be included which have shown a statistically significant impact on the examined strategic process constructs. If more than one variable in each group was found to be statistically significant, only variables that remained significant in the combined models are taken into consideration. Thus, the following team level antecedents of strategic decision making processes are selected: prior industry work experience, diversity of experiences, proportion of team members who belonged to the original founding team, team size, and gender distribution. Furthermore, life-cycle stage and technology orientation on the organisational level and effect uncertainty on the environmental level are included.

Figure 7.1 displays the whole model, and table 7.90 summarises the most important results. Prior industry work experience was found to have a statistically significant positive impact on the intensity of strategic planning activities, with a small effect size ($\beta = 0.190$, $p < 0.05$). The percentage of founders in the management team negatively and significantly influences planning, with a medium effect size ($\beta = -0.200$, $p < 0.05$). On the environmental level, effect uncertainty is statistically significantly and negatively associated with planning ($\beta = -0.254$, $p < 0.01$). All of the factors combined explain 22.4 per cent of the variance in planning. On the other hand, none of the predictors on the team level is statistically significantly related to the entrepreneurial orientation dimension of proactiveness. On the firm level, technology orientation positively influences the strategic orientation construct of proactiveness ($\beta = 0.460$, $p < 0.001$). The environmental factor effect uncertainty has a negative impact ($\beta = -0.189$, $p < 0.05$). Together, the factors explain 31.2 per cent of the total variance.

Planning and proactiveness each have a positive impact on the performance dimension of growth. The planning ($\beta = 0.340$, $p < 0.001$) and proactiveness ($\beta = 0.323$, $p < 0.001$) factors explain 25.1 per cent of the variance of the dependent variable. Most of the goodness-of-fit indices are in an appropriate range. The probability value of the chi-squared test of goodness-of-fit is not reached, and the NFI score is slightly below the recommended threshold. Figure 7.2 shows the full model including the strategic orientation variable of futurity instead of planning. Table 7.91 gives an overview of the main findings. On the team level, only the proportion of founders and the size of the executive team have statistically marginal impacts on long-term planning activities. The founders variable is negatively related ($\beta = -0.159$, $p <$

0.1) and team size positively related (β = 0.169, p < 0.1) to futurity. Whereas no statistically significant links could be identified on the organisation level, the perceived level of environmental uncertainty has a statistically significant impact (β = -0.228, p < 0.01) on the dependent variable. In sum, the predictors explain 16.7 per cent of the variance. Regarding the entrepreneurial orientation dimension of proactiveness, no significant relationships could be found on the team level. The technology orientation of the firm has a statistically significant positive effect (β = 0.459, p < 0.001), whereas effect uncertainty is negatively associated with proactiveness (β = -0.195, p < 0.05). The combination of factors explains 31.6 per cent of the overall variance.

Both futurity (β = 0.303, p < 0.001) and proactiveness (β = 0.335, p < 0.001) have statistically significant effects on firm performance. They explain 22.7 per cent of the variance of firm growth. Predominantly, the fit indices report a tolerable model fit except for the probability value of the chi-squared test of goodness-of-fit and the NFI score that is marginally not fulfilled.

Table 7.90: Integrated model of antecedents and consequences of strategic processes 1

	PLAN	PRO	GROWTH
Industry work experience	0.190*	0.018	
Diversity experience	0.046	0.115	
Founder	-0.200*	0.005	
Gender distribution	0.056	0.061	
Team size	0.053	-0.030	
Life-cycle stage	0.034	-0.087	
Effect uncertainty	-0.254**	-0.189*	
Technology orientation	0.095	0.460***	
Planning			0.340***
Proactiveness			0.323***
R^2	0.224	0.312	0.251
χ^2			124.1
df			97
P ($\geq$ 0.1)			0.033
χ^2/df ($\leq$ 3.0)			1.280
RMSEA ($\leq$ 0.1)			0.038
CFI ($\geq$ 0.9)			0.966
NFI ($\geq$ 0.9)			0.872
PCLOSE ($\geq$ 0.1)			0.841

[†]$p < 0.1$; *$p < 0.05$; **$p < 0.01$; ***$p < 0.001$

Figure 7.1: Integrated model of antecedents and consequences of strategic processes 1

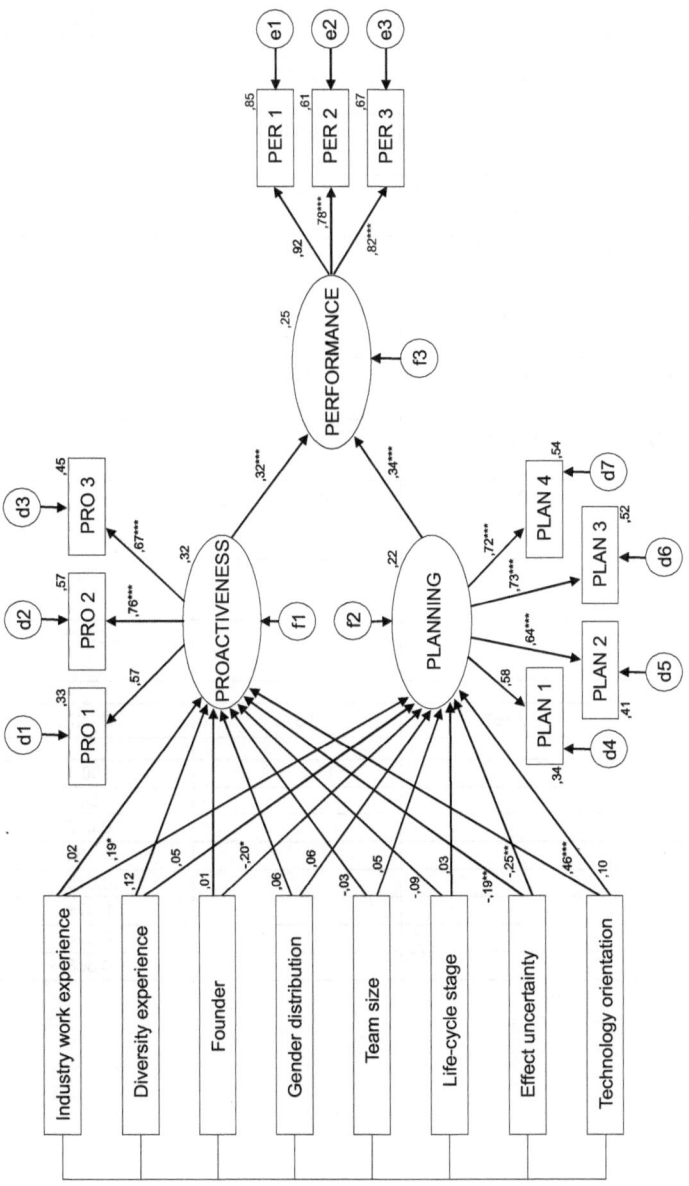

†p < 0.1; *p < 0.05; **p < 0.01; ***p < 0.001

Table 7.91: Integrated model of antecedents and consequences of strategic processes 2

	FUT	PRO	GROWTH
Industry work experience	0.072	0.023	
Diversity experience	0.008	0.114	
Founder	-0.159†	0.002	
Gender distribution	0.053	0.059	
Team size	0.169†	-0.032	
Life-cycle stage	0.023	-0.087	
Effect uncertainty	-0.228**	-0.195*	
Technology orientation	0.057	0.459***	
Futurity			0.303***
Proactiveness			0.335***
R^2	0.167	0.316	0.227
χ^2			100.3
df			81
P ($\geq$ 0.1)			0.072
χ^2/df ($\leq$ 3.0)			1.238
RMSEA ($\leq$ 0.1)			0.035
CFI ($\geq$ 0.9)			0.976
NFI ($\geq$ 0.9)			0.894
PCLOSE ($\geq$ 0.1)			0.868

†p < 0.1; *p < 0.05; **p < 0.01; ***p < 0.001

Figure 7.2: Integrated model of antecedents and consequences of strategic processes 2

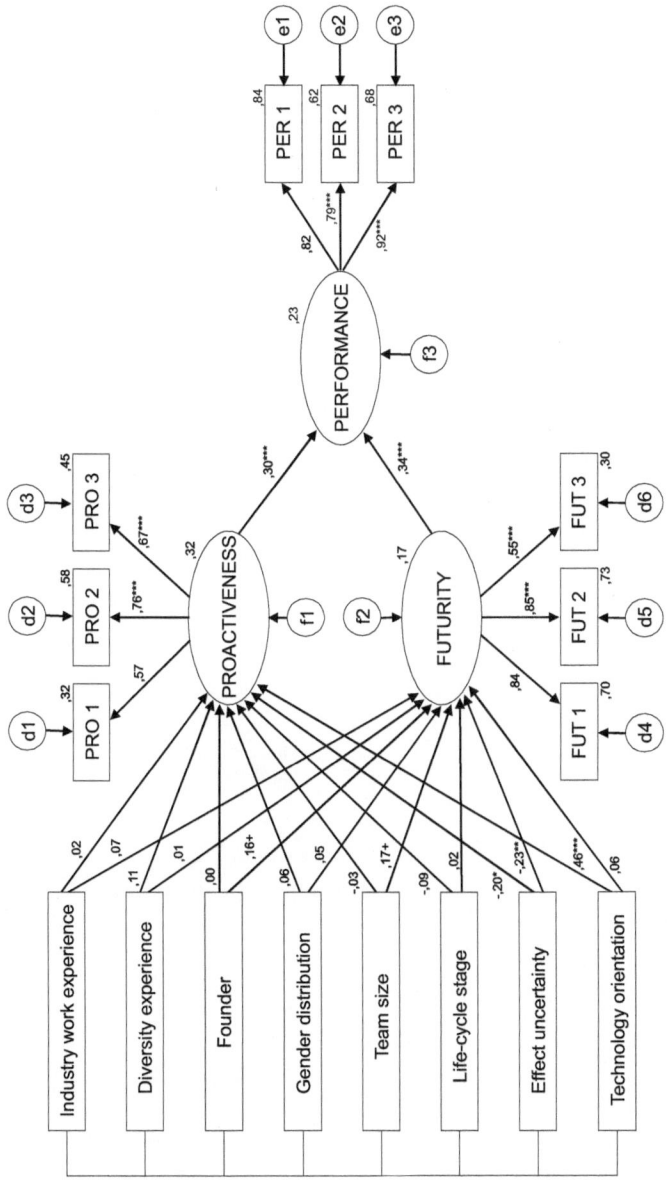

†p < 0.1; *p < 0.05; **p < 0.01; ***p < 0.001

8 Discussion

The starting point of this study was the question of how executives' characteristics influence strategic decision making and firm performance in the context of research-based spin-off companies. This final chapter summarises and discusses the findings. Furthermore, several implications for future research in the areas of academic entrepreneurship and strategic management are outlined. Possible implications for politicians, academic institutions, investors, and research-based spin-off companies are discussed. Finally, this chapter discusses the study's limitations and the most important conclusions that can be drawn from the study and the underlying empirical analysis.

8.1 Overview of the findings

The following section gives an overview of the results of the empirical analysis and briefly discusses each outcome. The first section deals with the relationship between management team characteristics and firm performance. A hypothesis is accepted if the factor has a statistically significant impact on at least one of the analysed performance dimensions: growth or profitability. In addition, the issues addressed by the research question are discussed. Second, the results regarding the link between strategic decision making and performance are described. In this section, the hypotheses consider formal and informal planning and each dimension of the entrepreneurial orientation concept. The third section discusses the results of the analysis of the relationships between executive team characteristics and strategic processes. Hypotheses were formulated for the expected impacts of prior professional experience and team diversity on formal and informal planning, respectively, and research questions were created for the remaining links. Finally, the findings regarding the factors on the firm and environmental levels are outlined.

8.1.1 Executives characteristics and firm performance

The analysis included firm growth and profitability as two salient measures of performance. The goodness-of-fit indices confirmed adequate convergent validity for both latent variables. The discriminant analysis clearly showed that growth and performance are distinct dimensions capturing different aspects of organisational performance. This finding is in agreement with previous research arguing that firm performance is a multidimensional phenomenon (Wiklund and Shepherd, 2003).

The results for the direct impact of prior professional experiences on firm performance only partly support the expected outcomes derived from the resource-based view and human capital theory. Prior entrepreneurial experience is only marginally significantly related to growth, so the first hypothesis in this subsection must be rejected. The findings of previous research on the relationship between prior entrepreneurial or business-ownership experience and firm performance are inconsistent. Experienced, so-called habitual or serial entrepreneurs do not self-evidently outperform inexperienced novice entrepreneurs (Ucbasaran, Westhead, and Wright, 2006). The former have likely gained more knowledge and capabilities, which should be beneficial for the creation and development of new ventures. However, they may also rely too much on their experience and invest less time in detailed planning or extensive information gathering. This behavioural pattern might limit the identification of opportunities, which is crucial for an emerging business in a changing environment. Because entrepreneurial and management teams are assumed to have an essential impact on spin-off success (Helm and Mauroner, 2007), inconsistent results call for a modified research design that also analyses how team characteristics and composition influence processes to identify mediating effects on firm performance.

Previous management experience in the team was found to have a statistically significant impact on growth but not on profitability. In contrast, prior industry work experience has a positive effect on profitability but is only marginally significantly related to growth. Thus, both hypotheses regarding the relationships between prior management and industry work experience and firm performance are accepted. Only the dichotomous variables capturing prior professional experience were found to be significantly related to performance. In fact, this result is also a limitation because the underlying measurement model does not exactly fulfil the requirements for the method, which assumes a metric measurement scale. However, the same model was run with a regression analysis and yielded very similar results, indicating the same significant links. Prior management and industry work experience are responsible for 3.2 and 3.7 per cent of the explained variance, respectively. According to the correlation matrix in table 7.1, the types of prior professional experience discussed here are related to one another. Unsurprisingly, previous entrepreneurial experience is often partnered with prior experience in a management position. The correlation between the two aspects is statistically significant ($r = 0.372$, $p < 0.01$). Management experience is also significantly correlated with prior industry work experience ($r = 0.296$, $p < 0.01$), but the correlation between entrepreneurial and industry work experience is not significant. Thus, these three variables capture different facets of prior experience, but they partially overlap. Future research should try to build on reflective constructs. For example, Chandler, Gaylen, and Hanks (1992, 1994) distinguished between the entrepreneurial, managerial, and technical functional roles that entrepreneurs can have in the venture and developed measurement scales for entrepreneurial and management competence that must be self-assessed on an individual level. As discussed above, this differentiation goes back to Penrose (1959).

Table 8.1: Prior professional experience and firm performance

Hypotheses	Outcome
Hypothesis 1.1.1: Prior entrepreneurial experience within the executive team has a positive impact on firm performance.	Rejected
Hypothesis 1.1.2: Prior management experience within the executive team has a positive impact on firm performance.	Accepted
Hypothesis 1.1.3: Prior industry work experience within the executive team has a positive impact on firm performance	Accepted

The influence of educational background was addressed by research questions. No significant relationships were observed. The accumulation of team members with a formal qualification in either a management- or technology-oriented field is not explicitly beneficial for the firm. The distribution of different educational backgrounds in the team is discussed later in this section. In particular, in the context of research-based spin-off companies, entrepreneurs and managers usually have high formal qualifications (Mosey and Wright, 2007). On average, almost two thirds of the team members hold a doctoral degree.

Table 8.2: Educational background and firm performance

Research questions	Outcome
Research question 1.1.1: How is the proportion of executives with an education in management related to firm performance?	No impact
Research question 1.1.2: How is the proportion of executives with an education in a technology-oriented field related to firm performance?	No impact
Research question 1.1.3: How is the overall level of education of the executive team related to firm performance?	No impact

Although many studies have emphasised the importance of a mix of competencies, knowledge, qualifications, and experience for a management team, only the diversity of the team's composition was found to be significantly related to growth. This relationship is of moderate strength and explains 5.0 per cent of the variance of the dependent variable. The results indicate that heterogeneity is not as crucial as previously believed. De Cleyn (2011) reported a positive impact of top management team heterogeneity in academic spin-offs in the early years after venture creation and showed that this effect diminishes over time. These findings might help explain the results of the empirical analysis, which did not exclusively focus on young companies in an early life-cycle stage.

Table 8.3: Heterogeneity and firm performance

Hypotheses	Outcome
Hypothesis 1.2.1: The degree of heterogeneity of the executive team in terms of prior professional experience has a positive impact on firm performance.	Rejected
Hypothesis 1.2.2: The degree of heterogeneity of the executive team in terms of educational background has a positive impact on firm performance	Rejected
Hypothesis 1.2.3: The degree of heterogeneity of the executive team in terms of general attributes has a positive impact on firm performance.	Accepted

Concerning the composition of executive teams in research-based spin-offs, prior research has discussed surrogate entrepreneurship as an important issue (e.g., Franklin, Wright, and Lockett, 2001; Vanaelst, Clarysse et al., 2006; Wright, Lockett et al., 2006). This term refers to external managers or entrepreneurs from outside the parent research organisation who join the executive team and complement or replace academic entrepreneurs who might lack sufficient managerial and entrepreneurial competencies. The empirical analysis has shown that, on average, less than one fifth of the team members are surrogate entrepreneurs. In addition, the data reveal that approximately one third of the spin-offs included a surrogate entrepreneur in their founding team when they launched the business. This fraction is equivalent to the proportion of external managers on the executive team. These figures indicate that management teams probably change much less, in terms of adding or exchanging team members, than prior studies have suggested (e.g., Vanaelst, Clarysse et al., 2006). Furthermore, the analysis shows no statistically significant direct impact of the proportion of external managers in the executive team on firm performance. Besides the main analysis, supplemental regressions were run using surrogate entrepreneurship as an indicator variable. None of these models reported any significant outcomes in terms of performance. However, these findings also raise the question of how the surrogate entrepreneurship phenomenon could be captured and measured appropriately. The explorative research questions addressed the potential influences of founders and owners, but no significant relationships could be identified.

Table 8.4: Composition and firm performance

Hypothesis and research questions	Outcome
Hypothesis 1.3: External managers in the executive team have a positive impact on firm performance.	Rejected
Research question 1.2.1: How is the proportion of founders in the management team related to firm performance?	No impact
Research question 1.2.2: How is the proportion of owners in the management team related to firm performance?	No impact

Of the examined general team characteristics, only the gender distribution was found to be significantly related to performance. According to the analysis, the proportion of males on the team has a positive impact on growth and profitability. On average, 92 per cent of the team members are males, and only 14 per cent of the spin-offs have at least one female manager on the executive team. The gender distribution of the data has little variance. Thus, these findings can only be discussed in the specific context of research-based spin-offs and cannot be generalised. Concerning the size of the management team, some scholars argue that large teams are beneficial for the firm because they possess more knowledge, experience, and capabilities (e.g., Bruton and Rubanik, 2002). De Cleyn (2011) observed that failed spin-offs tend to have smaller top management teams.

Table 8.5: General characteristics and firm performance

Research questions	Outcome
Research question 1.3.1: How is the size of the executive team related to firm performance?	No impact
Research question 1.3.2: How is the average age of the executive team related to firm performance?	No impact
Research question 1.3.3: How is the gender distribution in the executive team related to firm performance?	Positive impact

In summary, surprisingly few direct effects of management team characteristics on firm performance could be identified or confirmed given the attention paid to human capital-related issues in previous research. For example, the literature has often emphasised that venture capital decisions are based on the characteristics and composition of the entrepreneurial and management team rather than on the underlying technology or the written business plan.

8.1.2   Strategic decision making and firm performance

The second part of the empirical study examined the link between formal and informal approaches in terms of strategic decision making and firm performance. The results demonstrate a statistically significant positive relationship between the intensity of strategic planning activities and firm growth and a marginally significant impact on profitability. Long-term planning and analysis activities were found to have a significant positive impact on both growth and profitability. Compared to other predictors, the planning and futurity latent variables as single factors explain 13.8 and 11.3 per cent, respectively, of the variance of growth. They explain notably less of the profitability dependent variable, at 2.5 and 2.7 per cent, respectively. The fit of both measurement models was acceptable. One indicator of the futurity scale needed to be eliminated, and one item of the planning model scored a bit below the recommended benchmark for indicator reliability. However, the discriminant analysis demonstrated that the planning and futurity constructs are very similar. Nevertheless, both hypotheses addressing the effect of formal planning on organisational performance in the context of research-based spin-offs are accepted. In fact, the findings confirm the expected outcomes and merely replicate previous research in a new context. However, according to Hubbard, Vetter, and Little (1998), replication studies are also valuable because research should generally aim for empirical generalisation, which necessitates follow-up studies in different settings.

The analysis included two different aspects of informal strategy approaches. The level of flexibility was not significantly related to performance, but the latent partnering variable that reflects the degree of collaboration activities was found to have a positive impact on growth and profitability. There are two different interpretations for this particular finding. First, Chandler, DeTienne et al. (2011) showed that the indicators of this dimension, which they called *pre-commitments*, are related to the causation construct as well. As explained in the previous chapters, causation corresponds with the formal and rational approaches of strategic decision making. Thus, future research should discuss whether the *partnering* or *pre-commitments* construct (Chandler, DeTienne et al., 2011) can be seen solely as an informal and heuristic approach to strategic reasoning. Second, formal and informal strategising presumably do not contradict or exclude each other. Both approaches can affect firm performance positively or negatively. In addition, these effects may change over time when the venture grows and evolves through different life-cycle stages. Future research could address this particular issue with a longitudinal empirical study design. However, both latent variables discussed in this paragraph, flexibility and partnering, showed satisfactory goodness-of-fit indices, indicating appropriate convergent validity. The discriminant analysis confirmed that both aspects are salient dimensions.

Table 8.6: Formal and informal planning and firm performance

Hypotheses and research question	Outcome
Hypothesis 2.1: The level of formal strategic planning activities has a positive impact on firm performance.	Accepted
Hypothesis 2.2: The level of long-term firm orientation has a positive impact on firm performance.	Accepted
Research question 2: How does informal strategy making influence firm performance?	Partial positive impact

The dimensions of innovativeness, proactiveness, risk-taking, competitive aggressiveness, and autonomy were expected to be distinctive facets of the overarching entrepreneurial orientation concept. In fact, none of the conceptualised constructs fulfilled all of the required criteria for a good fit in terms of convergent validity of the measurement models. The models for the latent proactiveness and competitive aggressiveness variables were acceptable overall because most of the goodness-of-fit indices were satisfactory. However, the analysis of the innovativeness and risk-taking constructs yielded mixed results. The autonomy dimension even accounted for irritating outcomes and could not be considered for the empirical analysis. Intuitively, the innovativeness, proactiveness, competitive aggressiveness, and risk-taking dimensions seem to be distinguishable, but this assumption was not reflected in the underlying indicators and collected data. Presumably, the dimensions are too similar, as indicated by the discriminant analysis, which revealed that innovativeness and proactiveness in particular are not truly salient dimensions. Therefore, hypothesis 2.3 must be rejected. Future research must discuss more intensively whether entrepreneurial orientation as a firm-level behaviour is a unidimensional or multidimensional concept. In particular, scholars do not agree about whether autonomy is a dimension of the entrepreneurial orientation concept.

All entrepreneurial orientation dimensions were expected to have a positive impact on the performance of research-based spin-off companies, but only innovativeness, proactiveness, and competitive aggressiveness were significantly associated with growth. These results partly confirm previous research reporting a positive correlation between entrepreneurial orientation and firm performance (Rauch, Wiklund et al., 2009). Innovativeness also demonstrated a statistically marginal positive effect on profitability. However, the results for innovativeness must be discussed with caution given the poor fit of the underlying measurement model. The effect sizes and shares of the explained variance of the growth dependent variable are remarkably high, ranging between 9.4 and 16.0 per cent. The outcomes indicate that different aspects of the entrepreneurial orientation concept might vary regarding their influences on firm performance. A revised model with modified indicators and an acceptable model fit could probably clarify this particular issue. In reflective models, single indicators can be eliminated or changed to improve the overall fit of the measurement model. Therefore, in the next step, certain indicators should be removed or modified and the analyses repeated. Table 8.7 summarises the outcomes regarding the formulated hypotheses.

Table 8.7: Entrepreneurial strategic posture and firm performance

Hypotheses	Outcome
Hypothesis 2.3: Innovativeness, proactiveness, risk-taking, competitive aggressiveness, and autonomy are distinct dimensions of the entrepreneurial orientation concept.	Rejected
Hypothesis 2.4.1: The level of innovativeness is positively associated with firm performance.	Accepted
Hypothesis 2.4.2: The level of proactiveness is positively associated with firm performance.	Accepted
Hypothesis 2.4.3: The level of risk-taking is positively associated with firm performance.	Rejected
Hypothesis 2.4.4: The level of competitive aggressiveness is positively associated with firm performance.	Accepted
Hypothesis 2.4.5: The level of autonomy is positively associated with firm performance.	Excluded

8.1.3   Executive characteristics and strategic decision making

The novelty of this study lies in the link between the characteristics and compositions of the executive team and strategic decision making processes. Carpenter, Geletkancz, and Sanders (2004) noted the near absence of support for an impact of the demographic characteristics of the management team on strategic processes and postulated a need for a deeper understanding of how team characteristics influence organisational outcomes. The upper echelons framework has been criticised for being mostly a theoretical concept that remains a black box in many ways. This study sheds a small amount of light on the relationship between team characteristics and strategic decision making in the context of research-based spin-off companies. The results indicate that previous entrepreneurial, management, and industry work experience positively affect the level of strategic planning activities and the long-term orientation of the firm. On one hand, the findings correspond to the assumption that professional experience have an impact on strategic decision making, which confirms hypothesis 3.1.1. On the other hand, the direction of the effect is the opposite of the prediction, so hypotheses 3.1.2 and 3.1.3 must be rejected. At first, these results seem to contradict effectuation and causation theory, which state that experience and especially expertise should lead to a more effectual and heuristic approach to strategic reasoning and decision making. Effectuation and causation theory was introduced in the context of the venture creation process. The firms participating in this study are generally not extremely young, with an average age of 9.5 years. Obviously, this theory is not applicable to this particular context. Prior studies have applied effectuation and causation theory on the individual level, but this approach does not fit technology-oriented ventures, which are often run by a management team (Roberts, 1991a).

Research question 3.1 referred to the potential link between prior experience and the entrepreneurial orientation of the firm. The results indicate a statistically significant impact of prior entrepreneurial, management, and industry work experience on the dimension of competitive aggressiveness only, whereas the remaining dimensions, innovativeness, proactiveness, and risk-taking, are not affected.

The strength of this study is that it measures prior experience on the team level and connects it to outcomes on the organisational level. Gartner, Shaver et al. (1994) mentioned the need to study the entrepreneur on the team level rather than from an individual perspective. The entrepreneurial team has gained more attention in recent studies (e.g., Ensley, Pearson, and Pearce, 2003), but the measures face a number of limitations. First, they only reflect the executives' experience before they joined the company and do not consider their experience within the company. The impact of prior experience on current organisational outcomes might diminish as executives gain more experience within the observed company. Furthermore, the experience within the team is measured by several dichotomous variables that probably do not reflect the variance precisely enough in this context. Future research should try to capture previous experience on the team level in a more finely grained way. Westhead, Ucbasaran, and Wright (2005) discussed different aspects of entrepreneurial experience and distinguished between serial entrepreneurs, who have launched more than one company, and portfolio entrepreneurs, who own more than one company simultaneously.

Another interesting question would be whether prior experience with business failures or successes has different effects.

Table 8.8: Prior professional experience and strategic decision making

Hypotheses and research question	Outcome
Hypothesis 3.1.1: Previous professional experience of the executive team affects strategic planning activities.	Accepted
Hypothesis 3.1.2: Executive teams with prior entrepreneurial, management, or industry work experience emphasise less formal strategic planning activities.	Rejected
Hypothesis 3.1.3: Executive teams with prior entrepreneurial, management, or industry work experience emphasise more informal strategic planning activities.	Rejected
Research question 3.1: How does prior entrepreneurial, management, or industry work experience within the executive team influence the entrepreneurial strategic posture?	Positive impact on AGG

Research questions 3.2.1-3.2.3 focussed on the potential links between the educational backgrounds of the executive team members and strategic decision making. No statistically significant relationship could be found between education and formal planning activities. However, in terms of the discussed informal planning approaches, there was a significant link between a high proportion of managers with a technology-oriented education and the latent flexibility and partnering variables. The descriptive statistics reveal that, on average, almost 80 per cent of the spin-off management team members have a technology-oriented degree. According to the results, none of the entrepreneurial orientation dimensions were affected by the educational backgrounds of the management team members. Table 8.8 shows the outcomes for the discussed research questions.

An important question regarding the discussion of possible effects of varying educational backgrounds refers to the underlying measurement model. Scholars have applied a variety of approaches. In a quantitative empirical analysis of new technology-based ventures, Colombo and Grilli (2005) operationalised educational background by the average number of years of the founders' graduate and post-graduate education. They distinguished between managerial or economic and scientific or technical education and found positive links between both aspects and growth. Patzelt, zu Knyphausen-Aufseß, and Fischer (2009) analysed management teams of venture capital firms and measured education on the team level by the proportion of team members with a degree in management and science or engineering. They showed that management teams with differing educational backgrounds follow dissimilar investment strategies. In fact, educational background and decision making are linked. However, it is not clear what type of phenomenon is actually reflected by the different measurement models. Future research should try to develop a substantial theoretical explanation for the observed relationships.

Table 8.9: Educational background and strategic decision making

Research questions	Outcome
Research question 3.2.1: How does the educational background of the executive team influence formal strategic planning activities?	No impact
Research question 3.2.2: How does the educational background of the executive team influence informal strategic planning activities?	Positive impact of technology education
Research question 3.2.3: How does the educational background of the executive team influence the entrepreneurial strategic posture?	No impact

Besides the accumulation of certain experience or degrees, the distribution or mixture of different team characteristics was expected to strongly influence decision making behaviour. Hypotheses 3.2.2 and 3.2.3 stated that formal strategies are more likely to be applied by homogeneous teams, whereas diverse teams tend to be more informal in their decision making behaviour. The heterogeneity of prior professional experiences was found to have a statistically significant positive impact on strategic planning activities, which contradicts the predicted outcome. Furthermore, all of the analysed diversity aspects, including previous professional experience, educational background, and composition, have a significant positive impact on competitive aggressiveness. In the full model, 11.7 per cent of the total variance of the competitive aggressiveness dependent variable is explained. Thus, the findings support hypothesis 3.2.1 but not 3.2.2 and 3.2.3, which must be rejected. The outcomes of this section are summarised in table 8.10.

Notably, only one dimension of the entrepreneurial orientation concept was found to be significantly related to the diversity variables. The same observation has been made for the variables capturing prior professional experience.

Table 8.10: Heterogeneity and strategic decision making

Hypotheses and research question	Outcome
Hypothesis 3.2.1: Heterogeneity in the executive team affects the intensity of strategic planning activities.	Accepted
Hypothesis 3.2.2: Heterogeneous teams emphasise less formal strategic planning activities than homogeneous teams.	Rejected
Hypothesis 3.2.3: Heterogeneous teams emphasise more informal strategic planning activities than homogeneous teams.	Rejected
Research question 3.3: How does heterogeneity within the executive team influence the entrepreneurial strategic posture?	Positive impact on AGG

The relationships between the composition of the executive team and strategic decision making were addressed by research questions 3.4.1-3.4.3. Only the percentage of members of the current management team who belonged to the original founding team demonstrated a significant link to several aspects of strategy making. A high proportion of managers who have belonged to the company since incorporation (academic founding team) implies a lower level of formal planning activities. This relationship is statistically significant for both planning and futurity. Therefore, joint experiences within the team might reduce the need for extensive planning. Informal routines and knowledge about internal processes have probably been developed. This discussion partly reflects dynamics within the team, e.g., team member entry and exit. Future research should examine the proposition that more changes in the team lead to a higher level of strategic planning activities.

A high percentage of founders in the team also negatively affects partnering, which was originally defined as an informal strategy approach. This finding supports the previous discussion questioning this categorisation. Furthermore, the share of founders is negatively related to competitive aggressiveness, whereas none of the other entrepreneurial orientation dimensions is affected. This result is also in line with the findings discussed in the previous section. Table 8.11 shows the outcomes of the research questions.

Table 8.11: Composition and strategic decision making

Research questions	Outcome
Research question 3.4.1: How does the composition of the executive team influence formal strategic planning activities?	Negative impact of founders
Research question 3.4.2: How does the composition of the executive team influence informal strategic planning activities?	Negative impact of founders
Research question 3.4.3: How does the composition of the executive team influence the entrepreneurial strategic posture?	Negative impact of founders on AGG

Among the demographic characteristics and general attributes, the size of the management team is significantly related to planning and futurity and to competitive aggressiveness. The average age of the team members is negatively related to flexibility. A high proportion of males among the executives is positively associated with proactiveness. The outcomes of the research questions are displayed in table 8.12.

Table 8.12: General attributes and strategic decision making

Research questions	Outcome
Research question 3.5.1: How do the characteristics of the executive team influence formal strategic planning activities?	Positive impact of team size
Research question 3.5.2: How do the characteristics of the executive team influence informal strategic planning activities?	Negative impact of average age
Research question 3.5.3: How do the characteristics of the executive team influence the entrepreneurial strategic posture?	Positive impact of gender on PRO Positive impact of team size on AGG

8.1.4   Firm and environmental factors and strategic decision making

According to Vohora, Wright, and Lockett (2004), spin-off companies evolve through several distinct phases. During this development, they face critical junctions and must acquire different resources and capabilities. Therefore, a further stage in the life-cycle entails a higher degree of complexity and dynamics within the organisation. The more established the organisation is, the more routines could have already been developed, requiring less planning. The results reveal several statistically significant relationships between the explanatory variables reflecting several aspects of company maturity and the dependent variables that include different dimensions of strategic orientation on the firm level. This outcome confirms the first hypothesis of this section. However, the outcomes do not form a clear picture. The firm size was found to be positively related to futurity and partnering, representing aspects of formal and informal planning, respectively. This finding is also in line with the results in the previous chapters questioning whether partnering can be categorised as a facet of informal strategy making. In addition, the distribution of the firm size variable was not within a tolerable range. According to the results, the firm's age has a positive impact on flexibility, contrary to the expected outcome. Consequently, both hypotheses regarding the links between the maturity of the company and formal or informal strategy approaches must be rejected. The relationships between the independent variables and the entrepreneurial orientation dimensions are also characterised by mixed results. On one hand, the life-cycle stage has a negative impact on proactiveness and risk-taking. On the other hand, firm size is positively associated with competitive aggressiveness.

Table 8.13: Company maturity and strategic decision making

Company maturity	Outcome
Hypothesis 4.1.1: The maturity of the company affects the intensity of strategic planning.	Accepted
Hypothesis 4.1.2: Firms in later development and life-cycle phases emphasise more formal strategic planning activities.	Rejected
Hypothesis 4.1.3: Firms in an early development and life-cycle phase emphasise more informal strategic planning activities.	Rejected
Research question 4.1: How does the maturity of the company influence the entrepreneurial strategic posture?	Mixed findings

The following section discusses the outcomes regarding the impact of environmental uncertainty on strategic orientation. As expected, environmental uncertainty is negatively associated with the level of strategic planning activities. Interestingly, only effect uncertainty, which refers to uncertainty about the impact of environmental changes on the company, was statistically significant. State uncertainty relates to the changes of certain environmental factors and is, according to the results of the empirical analysis, positively related to flexibility. Therefore, the formulated hypotheses addressing the links between environmental uncertainty and formal and informal planning activities are accepted. However, the entrepreneurial orientation dimension of innovativeness was positively affected by the state uncertainty variable, whereas proactiveness was negatively associated with effect uncertainty. As shown in the discriminant analysis, the innovativeness and proactiveness dimensions are very similar. Therefore, the outcomes regarding the effect of perceived environmental uncertainty are noteworthy and indicate that the measurement models capturing environmental uncertainty must be discussed in more detail.

Previous studies analysing the effects of environmental factors usually distinguish between environmental dynamism, munificence, and uncertainty. Based on McKelvie, Haynie, and Gustavsson (2009), this study focuses on state and effect uncertainty as different types of uncertainty. Further research should also examine the effect of response uncertainty, which reflects the inability to predict possible responses to environmental changes on the firm level.

Table 8.14: Environmental uncertainty and strategic decision making

Hypotheses and research question	Outcome
Hypothesis 4.2.1: The perceived level of environmental uncertainty affects the intensity of strategic planning activities.	Accepted
Hypothesis 4.2.2: Firms in an uncertain environment emphasise less formal strategic planning activities.	Accepted
Hypothesis 4.2.3: Firms in an uncertain environment emphasise more informal strategic planning activities.	Accepted
Research question 4.2: How does the perceived level of environmental uncertainty influence the entrepreneurial strategic posture?	Mixed findings

Because the level of technology is an important aspect of research-based spin-off companies, the link to the strategic orientation of the firm was addressed by research questions 4.3.1-4.3.3. Interestingly, the results follow a clear pattern. On one hand, only the technology orientation variable is significantly related to partnering. On the other hand, the technology orientation, research intensity, and product focus explanatory variables have a statistically significant impact on almost all entrepreneurial orientation dimensions. Only the relationship between product focus and innovativeness is not significant.

Furthermore, this analysis revealed the largest effect sizes and explained variances in the study. Technology orientation explains 25.0 per cent of the variance of proactiveness, and in the full model, all factors explain 30.0 per cent. Research intensity explains 22.4 per cent of the variance of risk-taking, and in the full model, all independent variables remain statistically significant and are responsible for 33.7 per cent of the variance.

Previous research has discussed factors that are related to the technology base as possible predictors of firm performance (e.g., Riesenhuber, 2008) or moderators of the link between entrepreneurial orientation and economic outcomes. According to the upper echelons framework, this study considered the technology-related variables to directly affect the strategic orientation and, hence, have a mediating effect on performance. These results are in line with those of Borch, Huse, and Senneseth (1991), who found that technological firms tend to apply growth- and product-oriented strategies.

Table 8.15: Level of technology and strategic decision making

Research questions	Outcome
Research question 4.3.1: How does the technology base influence the level of formal strategic planning activities?	No impact
Research question 4.3.2: How does the technology base influence the level of informal strategic planning activities?	Positive impact on PART
Research question 4.3.3: How does the technology base influence the entrepreneurial strategic posture?	Positive impact

8.2   Implications

The following section outlines implications of this study. First, implications for scholars in the fields of academic entrepreneurship, strategic management, and strategic entrepreneurship are discussed. Second, implications for practitioners such as politicians, technology transfer officers, potential investors, and entrepreneurs and managers of research-based spin-off companies are briefly outlined.

8.2.1   Implications for researchers

This research must be positioned in both the entrepreneurship and the strategic management literatures. First, some general outcomes must be mentioned. The discriminant analysis revealed that growth and profitability are two distinct dimensions of firm performance. This finding confirms prior research demonstrating the multidimensionality of the performance construct (e.g., Chandler, and Hanks, 1993; Davidsson and Wiklund, 2003). Whereas previous studies have often used composite scales for firm performance, this analysis applied constructs with reflective indicators. As shown in chapter 7, the goodness-of-fit indices for the underlying measurement models reported satisfactory outcomes, indicating the validity and reliability of the latent variables. This insight is especially relevant for scholars performing quantitative empirical research on young ventures and small and medium-sized companies. On one hand, external sources, such as company databases or public registers, often contain little available information about these firm types. On the other hand, detailed information about revenue streams, profit, and other indicators might be too sensitive for the firms to disclose in a survey. Furthermore, some figures, such as the growth of very young ventures, might not reflect the situation properly because they tend to score highly, especially in the early stages of a company's life-cycle. Although it might be more favourable if the results referred to real performance figures, collecting self-perceived estimates of the company's current situation compared with that of close competitors is a practicable and feasible alternative when other data are unavailable.

Second, some implications can be drawn from the analysis of the relationship between executive team characteristics and the performance of academic spin-offs. This link refers to the discussion of success factors, especially in the context of new technology-based ventures and research-based spin-off companies in particular. As outlined in chapter 2, scholars have applied different definitions of an academic spin-off, which hinders the comparison of the study outcomes. In addition, no available public database gives an overview of the entire population. For this reason, before conducting the main survey, a preliminary study was performed in 24 European countries to delineate the boundaries and characteristics of the spin-off population in Europe. This study was one of the first research efforts to address the spin-off phenomenon on such a large scale with respect to both the number of countries included and the number of research institutes contacted. The relatively high response rate, yielding a database with more than 8,500 spin-off companies, and the additional information gained from corporate websites and external sources allowed an accurate estimation of the diffusion and characteristics of academic spin-offs in the European context.

Nevertheless, as outlined in chapter 5, representativeness is still a critical issue for academic entrepreneurship studies. Based on the underlying definition for this analysis, many companies needed to be excluded from the sample selection process. Therefore, some differences between the sample and the original population could not be avoided. As most quantitative empirical studies on research-based spin-offs do not refer to an entire population, the findings are actually problematic in terms of generalisability. Furthermore, empirical studies often over-interpret their own results, especially if the observed effects are very small. Therefore, the study should be replicated in a different context to provide evidence for or against the outcomes and compensate for the lack of representativeness.

According to the results of the empirical analysis, in many ways, this study could not confirm a direct impact of human capital of the management team on firm performance. This finding might enhance the contradictory discussions among scholars· about the impact of executives on organisational outcomes. For example, Unger, Rauch et al. (2011) included 70 independent samples in a meta-analysis and found a significant but small relationship between human capital and entrepreneurial success. Cannella, Finkelstein, and Hambrick (2008) asked the provoking question "Do executives matter?" and examined research on strategic leadership to understand the effects of top managers on strategy and performance. The essence of their comprehensive overview of recent studies is that executives play an important role for organisations, but the influence of human factors must be discussed in an integral way. Besides a direct impact, managers have an essential influence on the firm's behaviour, which might result in an indirect link to firm performance. As many studies on success factors in the academic entrepreneurship field refer to the resource-based view as a theoretical framework, the extension of this perspective to explain potential mediating effects should be promising. The upper echelons theory that emerged from the strategic management field provides a framework that considers strategy-related aspects and is also applicable in the entrepreneurship context.

Regarding the direct link between management team characteristics and firm performance, only a few variables were found to have a statistically significant impact, such as management and industry work experience. It is noteworthy that entrepreneurial experience was not significant. In this context, recent studies have discussed several facets of entrepreneurial experience. This stream of research differentiates between failed and successful entrepreneurs and defines those with experience in one or more prior companies as serial entrepreneurs. The main conclusion of these studies is that the types of previous entrepreneurial experience have varying effects on organisational outcomes. Thus, it is probably not sufficient to subsume all of these facets into one variable, as was conceptualised in this study. It is also noteworthy that only one facet of team diversity was found to be significantly related to performance. Neither experiential nor educational heterogeneity was significant; only composition played a role in this respect. External managers who are surrogate entrepreneurs have no statistically significant effect on growth or profitability which is another interesting outcome of the empirical analysis given the attention paid to the phenomenon of surrogate entrepreneurship in academic spin-offs (e.g., Radosevich, 1995; Franklin, Wright, and Lockett, 2001).

Implications of the findings regarding the relationships between different dimensions of strategic orientation on the firm level and performance are discussed in the following paragraphs. The conceptual framework underlying this study distinguished between formal and informal approaches to strategic decision making. Formal strategy, following the dominant logic of rationality in the decision making process, was represented by the two latent variables, planning and futurity, that were both derived from previous research in the fields of entrepreneurship and strategic management. According to the goodness-of-fit indices, both constructs are in an acceptable range, indicating sufficient validity and reliability. However, one item of the futurity construct needed to be excluded, and one indicator of the planning scale scored slightly below the recommended threshold. The discriminant analysis revealed that the latent planning and futurity variables are very similar and do not really reflect distinct phenomena of formal planning. This finding stimulates the discussion about the multidimensionality of strategic planning. As Podsakoff, Shen, and Podsakoff (2006) noted, many quantitative empirical studies in strategic management contained misspecifications of the measurement models. In most of these cases, multidimensional constructs were conceptualised as reflective measures, although the content and character of the item structure would require a formative measurement model. However, future studies must address this issue and develop reflective multi-item measurement models for distinct dimensions of strategic planning.

The latent flexibility, experimentation, and partnering variables represented different aspects of informal strategy and were also derived from prior research. As shown by the goodness-of-fit indices, the experimentation construct was insufficient in terms of validity and reliability and could not be used for the analysis. The latent flexibility and partnering variables demonstrated fit indices within an acceptable range, although two out of the four items of the flexibility construct needed to be removed from the model. The discriminant analysis verified that the flexibility and partnering variables represent different phenomena. However, as outlined in the previous sections, it is unclear whether partnering can be classified as a typical dimension of informal strategic decision making. In addition, Chandler, DeTienne et al. (2011) concluded that this dimension is related to both effectuation and causation and, thus, is simultaneously associated with rational and heuristic decision making. The results of the empirical analysis confirm that planning and futurity positively affect performance, in agreement with previous research. In addition, partnering has a positive effect on performance. The effect sizes are generally higher for the performance dimension of growth. The planning and futurity variables are responsible for 13.6 and 11.4 per cent of the variance of growth, respectively, whereas partnering explains only 5.5 per cent.

The goodness-of-fit indices for the measurement models of the entrepreneurial orientation dimensions demonstrated mixed results. None of the latent variables had satisfactory outcomes. Whereas the proactiveness and competitive aggressiveness dimensions were within a tolerable range, the innovativeness and risk-taking constructs showed questionable validity and reliability. In addition, the autonomy dimension could not be used for the analysis because it showed inconclusive fit indices. The discriminant analysis revealed that the entrepreneurial orientation dimensions are not all distinct from each other. In particular, innovativeness and proactiveness are very similar and, thus, reflect almost the same phenomenon. This outcome contributes to the discussion about the unidimensionality and multidimensionality of the entrepreneurial orientation scales. Scholars also must discuss more extensively which dimensions belong to a firm's entrepreneurial orientation. Whereas the dimensions of innovativeness, proactiveness, and risk-taking come from the original conceptualisation by Miller (1983), competitive aggressiveness and autonomy were added to the entrepreneurial orientation concept later (Lumpkin and Dess, 1996). For instance, most quantitative empirical studies do not consider autonomy (Rauch, Wiklund et al., 2009)[16]. In fact, not all entrepreneurial orientation dimensions are positively associated with performance; only proactiveness, innovativeness, and competitive aggressiveness have a statistically significant impact on growth. However, each factor explains a relatively high percentage of the variance. Innovativeness accounts for 15.8 per cent, proactiveness for 16.0 per cent, and competitive aggressiveness for 9.4 per cent. Previous studies have found that the entrepreneurial orientation dimensions contribute differently to firm performance. For instance, Dean (1993) noted that the competitive aggressiveness dimension explains more variace than the other facets of the entrepreneurial orientation concept.

In addition to outcomes of strategic decision processes, potential antecedents were examined as a central aspect of this study, with a focus on the composition and characteristics of the executive team. The empirical analysis disclosed several significant relationships between team characteristics and strategic decision making. However, the theory of

---

[16] More extensive discussion in Covin and Lumpkin (2011) and special issue of Entrepreneurship Theory and Practice 2011 on Entrepreneurial Orientation

effectuation and causation that was used to explain how expert entrepreneurs influence strategic decision making processes was definitely not applicable. For example, prior professional experience was found to have a statistically significant impact on formal planning, contrary to the hypotheses. The heterogeneity of prior experience is also positively related to planning. This outcome also does not correspond with the results expected under effectuation and causation theory. Several relationships were addressed by research questions because no substantial theoretical explanation could be given. The proportion of team members who belonged to the original founding team has a negative impact on the intensity of formal planning; team size is positively associated with the planning and futurity variables; and a focus on technology-oriented formal qualifications in the team is positively related to informal strategy. The findings contribute to the entrepreneurship and strategic management fields as Carpenter, Geletkancz, and Sanders (2004) noted that many aspects of the upper echelons framework remain poorly understood. However, the significant links identified must be backed by a conclusive theoretical framework. This study focuses only on directly observable characteristics, whereas the upper echelons framework also contains cognitive aspects. It will be very important to take these aspects into consideration within further studies, although the data collection procedure should be more sophisticated.

Regarding the impact of management team characteristics on a firm's entrepreneurial strategic posture, only the competitive aggressiveness dimension was found to be affected by several influence factors. Except for the link between gender distribution and proactiveness, no other relationship was found to be significant. According to the empirical analysis, experienced, diverse, and large teams tend to be more aggressive, whereas teams with a high proportion of original founders are less aggressive. As discussed in the previous paragraph, no theoretical framework can explain why the competitive aggressiveness dimension in particular is affected by several factors on the team level. Competitive aggressiveness was added to the entrepreneurial orientation construct by Lumpkin and Dess (1996). This finding should also encourage scholars in the field to clarify the specification and boundaries of the entrepreneurial orientation concept. The results could not confirm prior research showing a positive relationship between professionals, such as engineers and scientists, and the level of innovation in the company (e.g., Hage, 1980; Miller and Friesen, 1982).

The fourth link analysed in this study refers to factors on the organisational and environmental levels. In line with previous research and the formulated hypotheses, environmental uncertainty was found to have a statistically significant negative effect on formal planning activities. Two different types of environmental uncertainty were distinguished. Interestingly, only the effect uncertainty factor demonstrated a significant link to planning and futurity, with medium effect sizes explaining 9.9 per cent and 7.0 per cent of the variance, respectively. This finding shows that environmental uncertainty plays an important role in strategic decision making processes. The measurement models were derived from McKelvie, Haynie, and Gustavsson (2011), who applied different types of uncertainty in a simulation study. The empirical analysis confirms that different types of perceived environmental uncertainty have different effects on the behaviour of management teams. State uncertainty refers to the assumption of the general environment without a concrete connection to the firm and was found to have a statistically significant positive impact on informal strategy, reflected by the flexibility variable. In summary, these findings encourage researchers to investigate the facets of environmental uncertainty and the potential effects on behavioural aspects in more detail.

The results do not show a clear pattern regarding the impact of the examined firm level variables, age, life-cycle stage, and firm size, on the observed strategic processes. Thus, the findings neither support nor contradict life-cycle theory. However, the empirical analysis presented a consistent picture regarding the link between the technology base and strategic decision making processes. The level of technology has a statistically significant influence on all dimensions of entrepreneurial orientation. Furthermore, the explained variances are relatively high. For example, the technology orientation variable is responsible for 25.0 per cent and 22.7 per cent of the variance in proactiveness and risk-taking, respectively. Scholars have included technology factors in the entrepreneurial orientation discussion (Rauch, Wiklund et al., 2009). Whereas most research has treated technology-related aspects as moderating factors (e.g., Lumpkin and Dess, 1996), this study referred to the upper echelons theory and assumed a direct effect of technology factors on entrepreneurial orientation. Therefore, future research must consider this point of view as well.

Finally, the integrated models included several antecedents of strategic decision processes and the performance dimension of growth as the outcome. In the first model, the factors explain 22.4 per cent of the variance of planning and 31.2 per cent of proactiveness. In the second model, these factors explain 16.7 per cent of the variance of futurity and 31.6 per cent of proactiveness. As discussed above, planning and futurity overlap to a great extent and, thus, were examined in different models. Planning and proactiveness together explain 25.1 per cent of the variance of growth, whereas futurity and proactiveness explain 22.7 per cent. Planning, proactiveness, and futurity all remain statistically significant in the full models. The empirical analysis proved that formal planning, represented by the latent planning and futurity variables, and proactiveness, as a dimension of the entrepreneurial orientation concept, play important roles and have a substantial impact on the performance dimension of growth in the context of research-based spin-off companies. However, the levels of formal planning activities and entrepreneurial orientation are affected by different factors. Planning is influenced by prior professional experience, the proportion of original founders in the executive team, and perceived environmental uncertainty. In contrast, proactiveness is influenced by the underlying technology base and environmental uncertainty.

8.2.2 Implications for practitioners

This study is also interesting for those in charge of evaluating, selecting, or changing members of the executive team (e.g., VC, boards, TTO, consulting, entrepreneurs). In this context, the central question is which factors enhance the performance of research-based spin-off companies. This question must be further differentiated. As discussed above, growth and profitability can be seen as distinct facets of firm performance.

For political decision makers, the promotion of economic development is of great importance. The preliminary study confirmed that academic entrepreneurship is becoming increasingly relevant. In this context, it is crucial to know whether support schemes, especially for new technology-based ventures and academic spin-offs, are useful. Evaluation studies do not always find a positive correlation between support mechanisms and the number of spin-offs created. Thus, the knowledge of relevant influence factors enables politicians to customise these support and funding programmes that promote the commercialisation of research results by spinning out new ventures. The configuration of the entrepreneurial and management teams is often a key issue. Several support schemes require diverse and experienced executive teams. According to the empirical results, the direct impact of team characteristics on performance is very limited. In contrast, strategic decision making processes such as planning and proactiveness have substantial effects, each explaining a considerable amount of the variance. Therefore, policy makers should try to develop subsidy mechanisms that support spin-offs by combining planning activities with an entrepreneurial, proactive posture. Support schemes should focus more on the venture creation and development process than on the initial configuration, e.g., of the entrepreneurial team.

The preliminary study also confirmed that research organisations in Europe have recognised the importance of research-based spin-offs as an alternative path in the technology transfer process. The empirical results provide some interesting insights for the management of research institutes and the departments that are involved in the commercialisation process, such as technology transfer offices. Spinning off new ventures from a research organisation often involves conflicting goals of the involved stakeholders. On one hand, research organisations aiming to become an entrepreneurial institute or an entrepreneurial university claim to have the mission of supporting the creation of new ventures based on research in their organisation. On the other hand, the spin-off process implies that academic staff leaves the parent organisation to establish the new company. For this reason, the initial configuration of the entrepreneurial and management team of the spin-off is a key issue from the viewpoint of the mother institute. The literature on academic entrepreneurship has discussed whether the involvement of external entrepreneurs, also known as surrogate entrepreneurs, may enhance the spin-off's chances for success and decrease the issue of losing academic staff to the new company. According to the empirical results of this study, external managers do not have a statistically significant effect on either firm performance or strategic decision making processes. However, the proportion of founders within the executive team who belonged to the original founding team has a significant negative impact on formal planning, partnering, and competitive aggressiveness. All of these factors are positively associated with firm performance. Thus, there is an indirect effect of original founders in the team on performance. Because academic spin-offs are usually initiated by the researchers, this finding supports the implementation of mechanisms that encourage scientists to create and start new businesses and allow them to return to the mother institute when they can be replaced by external executives, for instance. This approach has already been used in practice.

The implications for investors and other stakeholders of research-based spin-off companies that can be drawn from this study are very similar to those for parent institutions. As outlined in chapter 2, investors consider both the underlying technology base and the entrepreneurial team in their evaluations of a potential financial engagement in the venture, e.g., in the form of risk capital. However, the characteristics and composition of the entrepreneurial and management team play a prominent role in the investment decision. In this context, prior professional experience, and especially a conclusive mix of managerial and technical competencies, is an important factor. The empirical results of this study confirm a positive impact of prior management and industry work experience on firm performance. Nevertheless, all of the observed effect sizes are small, and the explained variance of a single factor is always below 4.0 per cent. Regarding the heterogeneity of the management team, only diversity of composition was found to be significantly associated with performance, with a medium effect size, explaining 5.0 per cent of the variance of the dependent variable. In contrast, strategic processes such as planning and proactiveness have a much greater impact on growth. The question is whether potential financers can develop investment schemes that are linked to strategic processes in the companies rather than evaluating predominantly directly observable characteristics of the entrepreneurs and managers.

Finally, the potential implications of the findings for entrepreneurial and management teams of research-based spin-off companies must be discussed. The initial question formulated to outline the scope of this study asked whether executive teams of academic spin-offs should plan strategically or act entrepreneurially. In fact, the results of the empirical analysis demonstrate that formal planning and several dimensions of the entrepreneurial orientation concept have a positive impact on growth. Furthermore, discriminant analysis proved that formal planning and entrepreneurial orientation are distinct aspects of a firm's strategic orientation. Thus, formal planning and entrepreneurial thinking do not contradict each other. In addition to possible outcomes or consequences of strategic decision making behaviour, this study examined potential antecedents, first on the team level and second on the organisational and environmental levels. The findings confirmed that the characteristics of the executive team have substantial effects on strategic orientation variables. Thus, executives must keep in mind that adding, replacing, or excluding team members may change the strategic orientation of the company. This consideration might be highly relevant when experienced external managers, or surrogate entrepreneurs, join the team and replace academic entrepreneurs. As shown in the previous chapter, experienced, diverse, and larger teams emphasise formal planning activities to a greater extent and tend to be more aggressive. This difference in style may cause conflicts with the original founders, who most likely have a different strategic mind-set. The empirical results show that a high proportion of team members who belonged to the original founding team is negatively related to formal planning and competitive aggressiveness.

8.3 Limitations

Due to the methodological approach, the quantitative empirical analysis has several limitations that are outlined in the following section. First, concerns about the proposed causality must be mentioned. Studies applying the upper echelons framework often find significant correlations between observable characteristics of the management team and organisational outcomes such as performance, and they tend to argue that the causality goes from team characteristics to performance. However, it must be considered that the causality of these relationships could also be the other way around. For instance, Hellerstedt (2009) concluded that past performance influences team member entries, changing the composition and thus the characteristics of the team. The underlying theoretical framework and research model must plausibly explain the direction of the postulated effects between antecedents and outcomes. Research on upper echelon characteristics sometimes shows significant relationships between upper echelon characteristics and performance, but researchers wish to argue that differences in upper echelon characteristics cause differences in performance.

Second, the data collection constitutes a similar limitation. The empirical analysis is based on cross-sectional data that provide information on the variables of the observed companies at a given point of time. The main problem with this type of data involves the differentiation of causes and effects. If possible, future studies should observe changes in academic spin-offs' executive team, strategic orientation, and performance over time and build up a longitudinal data set that is not constrained in terms of the discussed causality problem.

Third, the results regarding the relationship between strategic processes and firm performance are probably constrained by common-method bias because dependent and independent variables were collected from the same source and both reflect estimations of the respondents that could not be validated by external sources. Multiple answers from 38 companies were used to limit common-method bias. t-tests comparing each company's first and second answers did not reveal significant differences ($p < 0.05$). However, future research could apply a design that includes performance measures based on real revenue figures.

Finally, limitations in terms of representativeness have been mentioned in the chapter 5. Because the companies of the original population were checked for whether they meet the underlying definition of a research-based spin-off company, several ventures needed to be excluded and could not be considered in the empirical analysis. This criterion resulted in a selection bias because not all members of the population are equally represented in the final sample. Additionally, failed spin-off companies could not be examined. Thus, the sample is also constrained by a survivorship bias as a special type of selection bias that can lead to overly optimistic interpretations of the findings.

8.4 Summary and final conclusions

The main objective of this study was to examine how executive team characteristics affect organisational outcomes such as strategy and performance. The context of research-based spin-off companies was selected because as a special subgroup of new technology-based ventures academic spin-offs face specific constraints such as liabilities of newness, size, or uncertainty. Furthermore, they rely on knowledge and technology transferred from the mother institute to the new venture and are usually established and managed by former academic staff members, especially in the early years of venture development. Therefore, entrepreneurial teams of research-based spin-offs often lack entrepreneurial knowledge and management skills and tend to be more homogeneous and less effective than teams in other traditional start-ups.

The link between human capital aspects and firm performance has often been examined through the theoretical lense of the resource-based view of the firm. This approach has several limitations and shortcomings because human capital attributes are generally considered as resources, although certain characteristics could be constraints for the venture development as well. For this reason, this study extended the theoretical background and connected the entrepreneurship literature with the strategic management field. The research stream of strategic entrepreneurship deals with overlapping topics and complementary contributions because both entrepreneurship and strategic management emphasise firm performance as a central aspect, but have different angles.

The research model considered characteristics of the executive team in academic spin-offs and linked them to strategic processes and firm performance. The study applied the upper echelons perspective as central theoretical framework and combined it with additional theoretical approaches including the resource-based view, human capital theory, effectuation and causation theory, stage-based models, and life-cycle theory. The upper echelons theory postulates that the characteristics of top managers have great impact on strategic decisions and organisational outcomes. The study design was complemented by certain factors on firm and environmental level such as technology base or level of uncertainty. Thus, the following relationships were analysed: (1) the link between executive characteristics and firm performance, (2) the link between strategic decision making and firm performance, (3) the link between executive characteristics and strategic decision making, and (4) the link between organisational and environmental aspects and strategic decision making.

The quantitative empirical analysis consisted of a preliminary study and a main survey. Because no external sources could provide a sufficient overview of the entire spin-off population, the preliminary study addressed the technology transfer offices of 809 research organisations in 24 European countries and collected data of more than 8,500 spin-off companies. The data was complemented by extensive desk research on corporate websites, business registers, and external data bases of company profiles. Descriptive statistics characterised and delineated the boundaries of the spin-off phenomenon in the European context. The main survey was conducted to examine the formulated hypotheses and research questions and sent to the managing directors of a subsample of academic spin-offs located in Germany. The survey included two reminders and personal telephone calls to all non-respondents, so an effective response rate of 31.64 per cent could be achieved based on the final data set with 193 spin-off companies.

The results reported only a few statistically significant relationships between executive team characteristics and firm performance. Prior management and industry work experience, diversity of team composition, and gender distribution were found to have direct positive effects on performance. This finding was surprising with regard to the extensive discussion about human capital aspects, especially in the context of small and medium-sized firms and technology-based ventures in particular. The second part of the analysis confirmed previous research in the field of strategic management. Strategic planning and long-term firm orientation were found to have a positive impact on performance. Furthermore, the dimensions innovativeness, proactiveness, and competitive aggressiveness of the entrepreneurial orientation concept were positively related to performance. The third link of this study dealt with the link between executive team characteristics and strategic processes on organisational level. In line with the upper echelons theory, several attributes of spin-off managers were found to have a statistically significant effect on strategic decision making. However, some outcomes turned out to have an opposite influence than expected. Prior professional work experience and heterogeneity in terms of former experience were positively associated with the level of strategic planning activities. In terms of entrepreneurial strategy making, only competitive aggressiveness was influenced by several factors. Finally, examining firm and environmental aspects disclosed a positive impact of technology base on all entrepreneurial orientation facets and a negative effect of perceived uncertainty on strategic planning.

In summary, the study contributed to existing literature in several ways. First, it applied the upper echelons theory to research-based spin-off companies focussing on small and young high technology ventures rather than on large and established corporations. Secondly, it contributed to the strategic entrepreneurship field by examining the link between executive team characteristics and strategic processes and firm performance. Finally, it empirically tested effectuation and causation theory in the context of strategic decision making. Although the expected outcomes could not be confirmed, the findings demonstrated that executive team characteristics have a substantial impact on strategic decision processes on organisational level. This sheds a small amount of light on the big black box of upper echelons theory.

Appendices

Additional tables:

General attributes and growth

	Model 1	Model 2	Model 3	Model 4
Team size	0.192*			0.190*
Average age		0.131		0.104
Gender distribution			0.186*	0.183*
R^2	0.037	0.017	0.035	0.083
χ^2	0.2	9.2	0.8	10.1
df	2	2	2	6
P ($\geq$ 0.1)	0.903	0.010	0.676	0.120
χ^2/df ($\leq$ 3.0)	0.102	4.581	0.391	1.687
RMSEA ($\leq$ 0.1)	0.000	0.137	0.000	0.060
CFI ($\geq$ 0.9)	1.000	0.975	1.000	0.986
NFI ($\geq$ 0.9)	0.999	0.969	1.000	0.967
PCLOSE ($\geq$ 0.1)	0.939	0.040	0.781	0.339

[†]$p < 0.1$; *$p < 0.05$; **$p < 0.01$; ***$p < 0.001$

General attributes and profitability

	Model 1	Model 2	Model 3	Model 4
Team size	0.029			0.020
Average age		0.077		0.068
Gender distribution			0.187[†]	0.186[†]
R^2	0.001	0.006	0.035	0.041
χ^2				1.7
df	0	0	0	2
P ($\geq$ 0.1)				0.420
χ^2/df ($\leq$ 3.0)				0.868
RMSEA ($\leq$ 0.1)				0.000
CFI ($\geq$ 0.9)				1.000
NFI ($\geq$ 0.9)				0.985
PCLOSE ($\geq$ 0.1)				0.573

[†]$p < 0.1$; *$p < 0.05$; **$p < 0.01$; ***$p < 0.001$

Professional experience and flexibility

	Model 1	Model 2	Model 3	Model 4
Entrepreneurial experience 1	-0.096			-0.114
Management experience 1		0.013		0.090
Industry work experience 1			-0.133	-0.150
R^2	0.009	0.000	0.018	0.032
χ^2				0.6
df	0	0	0	2
P ($\geq$ 0.1)				0.754
χ^2/df ($\leq$ 3.0)				0.283
RMSEA ($\leq$ 0.1)				0.000
CFI ($\geq$ 0.9)				1.000
NFI ($\geq$ 0.9)				0.996
PCLOSE ($\geq$ 0.1)				0.837

[†]$p < 0.1$; [*]$p < 0.05$; [**]$p < 0.01$; [***]$p < 0.001$

	Model 1	Model 2	Model 3	Model 4
Entrepreneurial experience 2	-0.084			-0.063
Management experience 2		-0.084		-0.014
Industry work experience 2			-0.121	-0.119
R^2	0.007	0.007	0.015	0.021
χ^2				0.1
df	0	0	0	2
P ($\geq$ 0.1)				0.938
χ^2/df ($\leq$ 3.0)				0.064
RMSEA ($\leq$ 0.1)				0.000
CFI ($\geq$ 0.9)				1.000
NFI ($\geq$ 0.9)				0.999
PCLOSE ($\geq$ 0.1)				0.961

[†]$p < 0.1$; [*]$p < 0.05$; [**]$p < 0.01$; [***]$p < 0.001$

Professional experience and partnering

	Model 1	Model 2	Model 3	Model 4
Entrepreneurial experience 1	0.024			0.006
Management experience 1		0.035		0.037
Industry work experience 1			0.118	0.036
R^2	0.001	0.001	0.014	0.002
χ^2				2.0
df	0	0	0	2
P (≥ 0.1)				0.366
χ^2/df (≤ 3.0)				1.004
RMSEA (≤ 0.1)				0.004
CFI (≥ 0.9)				0.989
NFI (≥ 0.9)				1.000
PCLOSE (≥ 0.1)				0.523

[†]$p < 0.1$; [*]$p < 0.05$; [**]$p < 0.01$; [***]$p < 0.001$

	Model 1	Model 2	Model 3	Model 4
Entrepreneurial experience 2	0.066			0.040
Management experience 2		0.022		0.024
Industry work experience 2			0.075	0.047
R^2	0.004	0.000	0.006	0.006
χ^2				1.2
df	0	0	0	2
P (≥ 0.1)				0.561
χ^2/df (≤ 3.0)				0.578
RMSEA (≤ 0.1)				0.000
CFI (≥ 0.9)				1.000
NFI (≥ 0.9)				0.994
PCLOSE (≥ 0.1)				0.693

[†]$p < 0.1$; [*]$p < 0.05$; [**]$p < 0.01$; [***]$p < 0.001$

Professional experience and proactiveness

	Model 1	Model 2	Model 3	Model 4
Entrepreneurial experience 1	0.109			0.077
Management experience 1		0.107		0.063
Industry work experience 1			0.077	0.049
R^2	0.012	0.012	0.006	0.018
χ^2	5.2	2.4	0.2	10,3
df	2	2	2	6
P (≥ 0.1)	0.075	0.300	0.918	0.114
χ^2/df (≤ 3.0)	2.591	1.205	0.086	1.711
RMSEA (≤ 0.1)	0.091	0.033	0.000	0.061
CFI (≥ 0.9)	0.969	0.996	1.000	0.970
NFI (≥ 0.9)	0.053	0.978	0.998	0.937
PCLOSE (≥ 0.1)	0.173	0.457	0.948	0.329

[†]$p < 0.1$; [*]$p < 0.05$; [**]$p < 0.01$; [***]$p < 0.001$

	Model 1	Model 2	Model 3	Model 4
Entrepreneurial experience 2	0.050			0.042
Management experience 2		0.048		0.018
Industry work experience 2			0.065	0.056
R^2	0.003	0.002	0.004	0.007
χ^2	4.8	1.3	0.1	6.8
df	2	2	2	6
P (≥ 0.1)	0.091	0.528	0.959	0.338
χ^2/df (≤ 3.0)	2.402	0.638	0.042	1.137
RMSEA (≤ 0.1)	0.085	0.000	0.000	0.027
CFI (≥ 0.9)	0.972	1.000	1.000	0.994
NFI (≥ 0.9)	0.956	0.988	0.999	0.955
PCLOSE (≥ 0.1)	0.198	0.666	0.975	0.610

[†]$p < 0.1$; [*]$p < 0.05$; [**]$p < 0.01$; [***]$p < 0.001$

Professional experience and innovativeness

	Model 1	Model 2	Model 3	Model 4
Entrepreneurial experience 1	0.064			0.037
Management experience 1		0.050		0.059
Industry work experience 1			-0.050	-0.075
R^2	0.004	0.002	0.002	0.009
χ^2				5.7
df	0	0	0	2
P ($\geq$ 0.1)				0.008
χ^2/df ($\leq$ 3.0)				2.874
RMSEA ($\leq$ 0.1)				0.099
CFI ($\geq$ 0.9)				0.874
NFI ($\geq$ 0.9)				0.844
PCLOSE ($\geq$ 0.1)				0.062

[†]$p < 0.1$; [*]$p < 0.05$; [**]$p < 0.01$; [***]$p < 0.001$

	Model 1	Model 2	Model 3	Model 4
Entrepreneurial experience 2	0.025			-0.058
Management experience 2		-0.035		-0.023
Industry work experience 2			-0.064	0.034
R^2	0.001	0.001	0.004	0.005
χ^2				3.8
df	0	0	0	2
P ($\geq$ 0.1)				0.073
χ^2/df ($\leq$ 3.0)				1.924
RMSEA ($\leq$ 0.1)				0.069
CFI ($\geq$ 0.9)				0.926
NFI ($\geq$ 0.9)				0.880
PCLOSE ($\geq$ 0.1)				0.251

[†]$p < 0.1$; [*]$p < 0.05$; [**]$p < 0.01$; [***]$p < 0.001$

Professional experience and risk-taking

	Model 1	Model 2	Model 3	Model 4
Entrepreneurial experience 1	-0.076			-0.086
Management experience 1		0.009		-0.003
Industry work experience 1			0.141	0.148
R^2	0.006	0.000	0.020	0.026
χ^2	5.1	0.01	1.3	7.3
df	2	2	2	6
P ($\geq$ 0.1)	0.078	0.995	0.531	0.297
χ^2/df ($\leq$ 3.0)	2.552	0.005	0.634	1.211
RMSEA ($\leq$ 0.1)	0.090	0.000	0.000	0.033
CFI ($\geq$ 0.9)	0.942	1.000	1.000	0.986
NFI ($\geq$ 0.9)	0.919	1.000	0.979	0.936
PCLOSE ($\geq$ 0.1)	0.178	0.997	0.668	0.569

[†]$p < 0.1$; [*]$p < 0.05$; [**]$p < 0.01$; [***]$p < 0.001$

	Model 1	Model 2	Model 3	Model 4
Entrepreneurial experience 2	-0.157[†]			-0.161[†]
Management experience 2		-0.014		-0.002
Industry work experience 2			0.091	0.098
R^2	0.025	0.000	0.008	0.034
χ^2	1.8	0.2	0.6	3.2
df	2	2	2	6
P ($\geq$ 0.1)	0.412	0.918	0.725	0.789
χ^2/df ($\leq$ 3.0)	0.886	0.086	0.322	0.527
RMSEA ($\leq$ 0.1)	0.000	0.000	0.000	0.000
CFI ($\geq$ 0.9)	1.000	1.000	1.000	1.000
NFI ($\geq$ 0.9)	0.971	0.997	0.989	0.969
PCLOSE ($\geq$ 0.1)	0.566	0.948	0.817	0.918

[†]$p < 0.1$; [*]$p < 0.05$; [**]$p < 0.01$; [***]$p < 0.001$

Education and planning

	Model 1	Model 2	Model 3	Model 4
Management education	0.084			0.151
Technology education		0.054		0.125
Level of education			0.015	0.018
R^2	0.007	0.003	0.000	0.020
χ^2	6.7	3.1	4.3	11.4
df	5	5	5	11
P ($\geq$ 0.1)	0.246	0.625	0.512	0.412
χ^2/df ($\leq$ 3.0)	1.334	0.681	0.853	1.035
RMSEA ($\leq$ 0.1)	0.042	0.000	0.000	0.013
CFI ($\geq$ 0.9)	0.990	1.000	1.000	0.998
NFI ($\geq$ 0.9)	0.965	0.983	0.977	0.958
PCLOSE ($\geq$ 0.1)	0.491	0.848	0.736	0.757

[†]p < 0.1; *p < 0.05; **p < 0.01; ***p < 0.001

Education and futurity

	Model 1	Model 2	Model 3	Model 4
Management education	0.091			0.062
Technology education		-0.065		-0.004
Level of education			-0.127	-0.110
R^2	0.008	0.004	0.016	0.020
χ^2	6.5	4.4	1.3	8.7
df	2	2	2	6
P ($\geq$ 0.1)	0.039	0.112	0.515	0.192
χ^2/df ($\leq$ 3.0)	3.255	2.190	0.663	1.446
RMSEA ($\leq$ 0.1)	0.108	0.079	0.000	0.048
CFI ($\geq$ 0.9)	0.976	0.987	1.000	0.990
NFI ($\geq$ 0.9)	0.967	0.978	0.993	0.969
PCLOSE ($\geq$ 0.1)	0.107	0.231	0.655	0.447

[†]p < 0.1; *p < 0.05; **p < 0.01; ***p < 0.001

Education and proactiveness

	Model 1	Model 2	Model 3	Model 4
Management education	0.001			-0.023
Technology education		-0.027		-0.027
Level of education			-0.045	-0.043
R^2	0.000	0.001	0.002	0.003
χ^2	0.7	0.2	0.8	1.9
df	2	2	2	6
P (≥ 0.1)	0.707	0.903	0.661	0.927
χ^2/df (≤ 3.0)	0.347	0.102	0.415	0.319
RMSEA (≤ 0.1)	0.000	0.000	0.000	0.000
CFI (≥ 0.9)	1.000	1.000	1.000	1.000
NFI (≥ 0.9)	0.993	0.998	0.992	0.989
PCLOSE (≥ 0.1)	0.804	0.939	0.770	0.976

[†]$p < 0.1$; *$p < 0.05$; **$p < 0.01$; ***$p < 0.001$

Education and innovativeness

	Model 1	Model 2	Model 3	Model 4
Management education	-0.004			0.007
Technology education		0.020		0.017
Level of education			0.006	0.004
R^2	0.000	0.000	0.000	0.000
χ^2				0.1
df	0	0	0	2
P (≥ 0.1)				0.941
χ^2/df (≤ 3.0)				0.061
RMSEA (≤ 0.1)				0.000
CFI (≥ 0.9)				1.000
NFI (≥ 0.9)				0.999
PCLOSE (≥ 0.1)				0.963

[†]$p < 0.1$; *$p < 0.05$; **$p < 0.01$; ***$p < 0.001$

Education and risk-taking

	Model 1	Model 2	Model 3	Model 4
Management education	-0.008			-0.001
Technology education		0.017		-0.010
Level of education			-0.004	0.020
R^2	0.000	0.000	0.000	0.000
χ^2	1.0	0.8	0.9	3.1
df	2	2	2	6
P ($\geq$ 0.1)	0.600	0.686	0.653	0.799
χ^2/df ($\leq$ 3.0)	0.510	0.377	0.427	0.513
RMSEA ($\leq$ 0.1)	0.000	0.000	0.000	0.000
CFI ($\geq$ 0.9)	1.000	1.000	1.000	1.000
NFI ($\geq$ 0.9)	0.983	0.987	0.985	0.977
PCLOSE ($\geq$ 0.1)	0.724	0.789	0.764	0.923

[†]$p < 0.1$; [*]$p < 0.05$; [**]$p < 0.01$; [***]$p < 0.001$

Education and competitive aggressiveness

	Model 1	Model 2	Model 3	Model 4
Management education	0.158			0.196
Technology education	.	-0.012		0.095
Level of education			-0.086	-0.045
R^2	0.025	0.000	0.007	0.033
χ^2				0.2
df	0	0	0	2
P ($\geq$ 0.1)				0.915
χ^2/df ($\leq$ 3.0)				0.088
RMSEA ($\leq$ 0.1)				0.000
CFI ($\geq$ 0.9)				1.000
NFI ($\geq$ 0.9)				0.999
PCLOSE ($\geq$ 0.1)				0.947

[†]$p < 0.1$; [*]$p < 0.05$; [**]$p < 0.01$; [***]$p < 0.001$

Heterogeneity and futurity

	Model 1	Model 2	Model 3	Model 4
Diversity experience	0.128			0.055
Diversity education		0.106		0.023
Diversity composition			0.186	$0.152^{\dagger}$
R^2	0.016	0.011	0.035	0.038
χ^2	0.3	0.3	1.5	1.7
df	2	2	2	6
P (≥ 0.1)	0.855	0.866	0.483	0.942
χ^2/df (≤ 3.0)	0.157	0.144	0.729	0.290
RMSEA (≤ 0.1)	0.000	0.000	0.000	0.000
CFI (≥ 0.9)	1.000	1.000	1.000	1.000
NFI (≥ 0.9)	0.998	0.999	0.993	0.994
PCLOSE (≥ 0.1)	0.907	0.914	0.628	0.982

†p < 0.1; *p < 0.05; **p < 0.01; ***p < 0.001

Heterogeneity and flexibility

	Model 1	Model 2	Model 3	Model 4
Diversity experience	-0.016			0.067
Diversity education		-0.066		-0.025
Diversity composition			-0.080	-0.039
R^2	0.000	0.004	0.006	0.004
χ^2				0.6
df	0	0	0	2
P (≥ 0.1)				0.725
χ^2/df (≤ 3.0)				0.322
RMSEA (≤ 0.1)				0.000
CFI (≥ 0.9)				1.000
NFI (≥ 0.9)				0.996
PCLOSE (≥ 0.1)				0.817

†p < 0.1; *p < 0.05; **p < 0.01; ***p < 0.001

Heterogeneity and partnering

	Model 1	Model 2	Model 3	Model 4
Diversity experience	0.020			0.051
Diversity education		-0.103		-0.145
Diversity composition			0.014	0.057
R^2	0.000	0.011	0.000	0.017
χ^2				0.6
df	0	0	0	2
P (≥ 0.1)				0.750
χ^2/df (≤ 3.0)				0.288
RMSEA (≤ 0.1)				0.000
CFI (≥ 0.9)				1.000
NFI (≥ 0.9)				0.997
PCLOSE (≥ 0.1)				0.835

[†]$p < 0.1$; [*]$p < 0.05$; [**]$p < 0.01$; [***]$p < 0.001$

Heterogeneity and innovativeness

	Model 1	Model 2	Model 3	Model 4
Diversity composition	0.156			0.079
Diversity education		0.053		0.134
Diversity experience			0.179	0.020
R^2	0.024	0.003	0.032	0.037
χ^2				0.5
df	0	0	0	2
P (≥ 0.1)				0.791
χ^2/df (≤ 3.0)				0.235
RMSEA (≤ 0.1)				0.000
CFI (≥ 0.9)				1.000
NFI (≥ 0.9)				0.996
PCLOSE (≥ 0.1)				0.864

[†]$p < 0.1$; [*]$p < 0.05$; [**]$p < 0.01$; [***]$p < 0.001$

Heterogeneity and proactiveness

	Model 1	Model 2	Model 3	Model 4
Diversity composition	0.140			0.071
Diversity education		$0.150^{\dagger}$		0.089
Diversity experience			$0.154^{\dagger}$	0.085
R^2	0.019	0.022	0.024	0.036
χ^2	1.8	0.8	0.8	5.4
df	2	2	2	6
P (≥ 0.1)	0.417	0.685	0.667	0.498
χ^2/df (≤ 3.0)	0.876	0.378	0.406	0.894
RMSEA (≤ 0.1)	0.000	0.000	0.000	0.000
CFI (≥ 0.9)	1.000	1.000	1.000	1.000
NFI (≥ 0.9)	0.984	0.993	0.993	0.973
PCLOSE (≥ 0.1)	0.570	0.788	0.774	0.744

$^{\dagger}p < 0.1$; $^{*}p < 0.05$; $^{**}p < 0.01$; $^{***}p < 0.001$

Heterogeneity and risk-taking

	Model 1	Model 2	Model 3	Model 4
Diversity composition	0.023			-0.074
Diversity education		0.133		0.094
Diversity experience			$0.161^{\dagger}$	0.150
R^2	0.001	0.018	0.026	0.034
χ^2	4.8	1.1	1.6	6.0
df	2	2	2	6
P (≥ 0.1)	0.089	0.567	0.447	0.418
χ^2/df (≤ 3.0)	2.417	0.568	0.805	1.008
RMSEA (≤ 0.1)	0.086	0.000	0.000	0.007
CFI (≥ 0.9)	0.946	1.000	1.000	1.000
NFI (≥ 0.9)	0.923	0.981	0.974	0.961
PCLOSE (≥ 0.1)	0.196	0.697	0.597	0.682

$^{\dagger}p < 0.1$; $^{*}p < 0.05$; $^{**}p < 0.01$; $^{***}p < 0.001$

Composition and proactiveness

	Model 1	Model 2	Model 3	Model 4
External managers	-0.044			-0.121
Ownership		-0.040		-0.046
Founder			-0.040	-0.099
R^2	0.002	0.002	0.002	0.011
χ^2	2.5	0.3	1.1	2.9
df	2	2	2	6
P (≥ 0.1)	0.284	0.844	0.566	0.823
χ^2/df (≤ 3.0)	1.259	0.170	0.568	0.481
RMSEA (≤ 0.1)	0.037	0.000	0.000	0.000
CFI (≥ 0.9)	0.995	1.000	1.000	1.000
NFI (≥ 0.9)	0.976	0.997	0.989	0.987
PCLOSE (≥ 0.1)	0.440	0.900	0.697	0.934

†p < 0.1; *p < 0.05; **p < 0.01; ***p < 0.001

Composition and innovativeness

	Model 1	Model 2	Model 3	Model 4
External managers	-0.012			-0.021
Ownership		-0.138		-0.011
Founder			-0.094	-0.005
R^2	0.000	0.019	0.009	0.000
χ^2				5.5
df	0	0	0	2
P (≥ 0.1)				0.063
χ^2/df (≤ 3.0)				2.758
RMSEA (≤ 0.1)				0.096
CFI (≥ 0.9)				0.975
NFI (≥ 0.9)				0.964
PCLOSE (≥ 0.1)				0.154

†p < 0.1; *p < 0.05; **p < 0.01; ***p < 0.001

Composition and risk-taking

	Model 1	Model 2	Model 3	Model 4
External managers	0.049			0.002
Ownership		-0.051		-0.032
Founder			-0.070	-0.059
R^2	0.002	0.003	0.005	0.006
χ^2	5.8	2.6	0.9	8.8
df	2	2	2	6
P ($\geq$ 0.1)	0.055	0.269	0.650	0.186
χ^2/df ($\leq$ 3.0)	2.895	1.312	0.430	1.465
RMSEA ($\leq$ 0.1)	0.099	0.040	0.000	0.049
CFI ($\geq$ 0.9)	0.929	0.988	1.000	0.983
NFI ($\geq$ 0.9)	0.909	0.957	0.985	0.953
PCLOSE ($\geq$ 0.1)	0.139	0.425	0.762	0.438

$^\dagger p < 0.1$; $^* p < 0.05$; $^{**} p < 0.01$; $^{***} p < 0.001$

General attributes and planning

	Model 1	Model 2	Model 3	Model 4
Team size	0.200*			0.202*
Average age		0.037		0.029
Gender distribution			0.086	0.090
R^2	0.040	0.001	0.007	0.049
χ^2	2.5	5.1	8.7	12.1
df	5	5	5	11
P ($\geq$ 0.1)	0.779	0.404	0.124	0.359
χ^2/df ($\leq$ 3.0)	0.497	1.020	1.731	1.097
RMSEA ($\leq$ 0.1)	0.000	0.010	0.062	0.022
CFI ($\geq$ 0.9)	1.000	0.999	0.979	0.985
NFI ($\geq$ 0.9)	0.987	0.973	0.955	0.941
PCLOSE ($\geq$ 0.1)	0.903	0.648	0.324	0.716

$^\dagger p < 0.1$; $^* p < 0.05$; $^{**} p < 0.01$; $^{***} p < 0.001$

General attributes and futurity

	Model 1	Model 2	Model 3	Model 4
Team size	0.264***			0.264***
Average age		-0.034		-0.034
Gender distribution			0.083	0.091
R^2	0.070	0.001	0.007	0.078
χ^2	1.0	6.2	0.7	7.8
df	2	2	2	6
P ($\geq$ 0.1)	0.608	0.045	0.718	0.256
χ^2/df ($\leq$ 3.0)	0.498	3.112	0.332	1.294
RMSEA ($\leq$ 0.1)	0.000	0.105	0.000	0.039
CFI ($\geq$ 0.9)	1.000	0.978	1.000	0.991
NFI ($\geq$ 0.9)	0.995	0.969	0.997	0.964
PCLOSE ($\geq$ 0.1)	0.730	0.119	0.812	0.525

[†]$p < 0.1$; *$p < 0.05$; **$p < 0.01$; ***$p < 0.001$

General attributes and flexibility

	Model 1	Model 2	Model 3	Model 4
Team size	0.130			0.168[†]
Average age		-0.277**		-0.265*
Gender distribution			-0.052	-0.040
R^2	0.017	0.077	0.003	0.097
χ^2				2.0
df	0	0	0	2
P ($\geq$ 0.1)				0.372
χ^2/df ($\leq$ 3.0)				0.988
RMSEA ($\leq$ 0.1)				0.000
CFI ($\geq$ 0.9)				1.000
NFI ($\geq$ 0.9)				0.979
PCLOSE ($\geq$ 0.1)				0.529

[†]$p < 0.1$; *$p < 0.05$; **$p < 0.01$; ***$p < 0.001$

General attributes and partnering

	Model 1	Model 2	Model 3	Model 4
Team size	0.099			0.114
Average age		-0.054		-0.116
Gender distribution			0.135	$0.131^{†}$
R^2	0.010	0.003	0.018	0.040
χ^2				2.7
df	0	0	0	2
P ($\geq$ 0.1)				0.258
χ^2/df ($\leq$ 3.0)				1.355
RMSEA ($\leq$ 0.1)				0.043
CFI ($\geq$ 0.9)				0.995
NFI ($\geq$ 0.9)				0.982
PCLOSE ($\geq$ 0.1)				0.413

$^{†}p < 0.1$; $^{*}p < 0.05$; $^{**}p < 0.01$; $^{***}p < 0.001$

General attributes and innovativeness

	Model 1	Model 2	Model 3	Model 4
Team size	0.192			$0.224^{†}$
Average age		-0.232		$-0.253^{†}$
Gender distribution			0.119	0.132
R^2	0.037	0.054	0.014	0.120
χ^2				0.2
df	0	0	0	2
P ($\geq$ 0.1)				0.097
χ^2/df ($\leq$ 3.0)				0.908
RMSEA ($\leq$ 0.1)				0.000
CFI ($\geq$ 0.9)				1.000
NFI ($\geq$ 0.9)				0.995
PCLOSE ($\geq$ 0.1)				0.942

$^{†}p < 0.1$; $^{*}p < 0.05$; $^{**}p < 0.01$; $^{***}p < 0.001$

General attributes and proactiveness

	Model 1	Model 2	Model 3	Model 4
Team size	0.070			0.078
Average age		-0.177		-0.191[†]
Gender distribution			0.190*	0.201*
R^2	0.005	0.031	0.036	0.076
χ^2	2.3	13.3	1.1	17.0
df	2	2	2	6
P ($\geq$ 0.1)	0.318	0.001	0.578	0.009
χ^2/df ($\leq$ 3.0)	1.147	6.654	0.547	2.829
RMSEA ($\leq$ 0.1)	0.028	0.172	0.000	0.098
CFI ($\geq$ 0.9)	0.997	0.897	1.000	0.900
NFI ($\geq$ 0.9)	0.979	0.889	0.990	0.870
PCLOSE ($\geq$ 0.1)	0.475	0.008	0.707	0.066

[†]$p < 0.1$; *$p < 0.05$; **$p < 0.01$; ***$p < 0.001$

General attributes and risk-taking

	Model 1	Model 2	Model 3	Model 4
Team size	0.073			0.074
Average age		-0.101		-0.120
Gender distribution			0.160[†]	0.162[†]
R^2	0.005	0.010	0.026	0.042
χ^2	5.4	1.9	4.3	11.3
df	2	2	2	6
P ($\geq$ 0.1)	0.068	0.389	0.115	0.079
χ^2/df ($\leq$ 3.0)	2.695	0.944	2.164	1.887
RMSEA ($\leq$ 0.1)	0.094	0.000	0.078	0.068
CFI ($\geq$ 0.9)	0.937	1.000	0.957	0.900
NFI ($\geq$ 0.9)	0.915	0.969	0.933	0.847
PCLOSE ($\geq$ 0.1)	0.161	0.544	0.235	0.263

[†]$p < 0.1$; *$p < 0.05$; **$p < 0.01$; ***$p < 0.001$

General attributes and competitive aggressiveness

	Model 1	Model 2	Model 3	Model 4
Team size	0.167**			0.214**
Average age		-0.080		-0.085
Gender distribution			0.161	$0.122^{\dagger}$
R^2	0.028	0.006	0.026	0.064
χ^2				1.3
df	0	0	0	2
P (≥ 0.1)				0.535
χ^2/df (≤ 3.0)				0.626
RMSEA (≤ 0.1)				0.000
CFI (≥ 0.9)				1.000
NFI (≥ 0.9)				0.982
PCLOSE (≥ 0.1)				0.672

$^{\dagger}p < 0.1$; $^*p < 0.05$; $^{**}p < 0.01$; $^{***}p < 0.001$

Company maturity and planning

	Model 1	Model 2	Model 3	Model 4
Firm age	-0.115			-0.193*
Life-cycle stage		0.036		0.018
Firm size (employees)			0.106	$0.198^\dagger$
R^2	0.013	0.001	0.011	0.049
χ^2	4.5	4.1	4.6	8.7
df	5	5	5	11
P ($\geq$ 0.1)	0.476	0.536	0.469	0.652
χ^2/df ($\leq$ 3.0)	0.905	0.819	0.916	0.789
RMSEA ($\leq$ 0.1)	0.000	0.000	0.000	0.000
CFI ($\geq$ 0.9)	1.000	1.000	1.000	1.000
NFI ($\geq$ 0.9)	0.976	0.978	0.976	0.968
PCLOSE ($\geq$ 0.1)	0.708	0.753	0.702	0.896

†p < 0.1; *p < 0.05; **p < 0.01; ***p < 0.001

Company maturity and futurity

	Model 1	Model 2	Model 3	Model 4
Firm age	0.017			-0.093
Life-cycle stage		0.040		-0.047
Firm size (employees)			0.310***	0.364***
R^2	0.000	0.002	0.096	0.111
χ^2	5.7	2.2	0.6	10.0
df	2	2	2	6
P ($\geq$ 0.1)	0.058	0.328	0.737	0.123
χ^2/df ($\leq$ 3.0)	2.855	1.115	0.305	1.672
RMSEA ($\leq$ 0.1)	0.098	0.024	0.000	0.059
CFI ($\geq$ 0.9)	0.980	0.999	1.000	0.985
NFI ($\geq$ 0.9)	0.971	0.989	0.997	0.965
PCLOSE ($\geq$ 0.1)	0.143	0.486	0.826	0.345

†p < 0.1; *p < 0.05; **p < 0.01; ***p < 0.001

Company maturity and partnering

	Model 1	Model 2	Model 3	Model 4
Firm age	0.037			-0.021
Life-cycle stage		-0.012		-0.053
Firm size (employees)			0.163*	0.241*
R^2	0.001	0.000	0.026	0.049
χ^2				1.0
df	0	0	0	2
P (≥ 0.1)				0.595
χ^2/df (≤ 3.0)				0.519
RMSEA (≤ 0.1)				0.000
CFI (≥ 0.9)				1.000
NFI (≥ 0.9)				0.995
PCLOSE (≥ 0.1)				0.720

$^\dagger p < 0.1$; $^*p < 0.05$; $^{**}p < 0.01$; $^{***}p < 0.001$

Company maturity and proactiveness

	Model 1	Model 2	Model 3	Model 4
Firm age	-0.007			0.082
Life-cycle stage		-0.193*		-0.307**
Firm size (employees)			0.057	0.178†
R^2	0.000	0.037	0.003	0.079
χ^2	0.7	3.7	0.3	3.8
df	2	2	2	6
P (≥ 0.1)	0.695	0.155	0.844	0.703
χ^2/df (≤ 3.0)	0.364	1.867	0.169	0.634
RMSEA (≤ 0.1)	0.000	0.067	0.000	0.000
CFI (≥ 0.9)	1.000	0.983	1.000	1.000
NFI (≥ 0.9)	0.993	0.967	0.997	0.980
PCLOSE (≥ 0.1)	0.795	0.290	0.900	0.874

$^\dagger p < 0.1$; $^*p < 0.05$; $^{**}p < 0.01$; $^{***}p < 0.001$

Company maturity and innovativeness

	Model 1	Model 2	Model 3	Model 4
Firm age	0.211			0.157
Life-cycle stage		$0.145^{\dagger}$		-0.034
Firm size (employees)			0.206	0.208
R^2	0.044	0.021	0.043	0.082
χ^2				1.6
df	0	0	0	2
P ($\geq$ 0.1)				0.449
χ^2/df ($\leq$ 3.0)				0.801
RMSEA ($\leq$ 0.1)				0.000
CFI ($\geq$ 0.9)				1.000
NFI ($\geq$ 0.9)				0.984
PCLOSE ($\geq$ 0.1)				0.599

$^{\dagger}p < 0.1$; $^{*}p < 0.05$; $^{**}p < 0.01$; $^{***}p < 0.001$

Company maturity and competitive aggressiveness

	Model 1	Model 2	Model 3	Model 4
Firm age	0.016			-0.136
Life-cycle stage		0.035		-0.043
Firm size (employees)			0.354^{***}	0.441^{***}
R^2	0.000	0.001	0.125	0.162
χ^2				2.1
df	0	0	0	2
P ($\geq$ 0.1)				0.343
χ^2/df ($\leq$ 3.0)				1.071
RMSEA ($\leq$ 0.1)				0.019
CFI ($\geq$ 0.9)				0.999
NFI ($\geq$ 0.9)				0.985
PCLOSE ($\geq$ 0.1)				0.500

$^{\dagger}p < 0.1$; $^{*}p < 0.05$; $^{**}p < 0.01$; $^{***}p < 0.001$

Environmental uncertainty and risk-taking

	Model 1	Model 2	Model 3	Model 4
State uncertainty	0.100			$0.185^{\dagger}$
Effect uncertainty		-0.151		-0.223*
R^2	0.010	0.023		0.051
χ^2	1.8	0.4		2.6
df	2	2		4
P (≥ 0.1)	0.409	0.826		0.631
χ^2/df (≤ 3.0)	0.893	0.191		0.644
RMSEA (≤ 0.1)	0.000	0.000		0.000
CFI (≥ 0.9)	1.000	1.000		1.000
NFI (≥ 0.9)	0.970	0.994		0.973
PCLOSE (≥ 0.1)	0.563	0.888		0.799

†p < 0.1; *p < 0.05; **p < 0.01; ***p < 0.001

Environmental uncertainty and competitive aggressiveness

	Model 1	Model 2	Model 3	Model 4
State uncertainty	0.019			0.011
Effect uncertainty		-0.036		-0.044
R^2	0.000	0.001		0.002
χ^2				0.2
df	0	0		1
P (≥ 0.1)				0.648
χ^2/df (≤ 3.0)				0.209
RMSEA (≤ 0.1)				0.000
CFI (≥ 0.9)				1.000
NFI (≥ 0.9)				0.998
PCLOSE (≥ 0.1)				0.718

†p < 0.1; *p < 0.05; **p < 0.01; ***p < 0.001

Technology base and planning

	Model 1	Model 2	Model 3	Model 4
Technology orientation	0.135			0.066
Research intensity		$0.146^{\dagger}$		0.093
Product focus			$0.141^{\dagger}$	0.089
R^2	0.018	0.021	0.020	0.036
χ^2	2.3	7.5	4.9	16.1
df	5	5	5	11
P ($\geq$ 0.1)	0.811	0.188	0.427	0.137
χ^2/df ($\leq$ 3.0)	0.454	1.493	0.982	1.464
RMSEA ($\leq$ 0.1)	0.000	0.051	0.000	0.049
CFI ($\geq$ 0.9)	1.000	0.986	1.000	0.979
NFI ($\geq$ 0.9)	0.988	0.961	0.974	0.942
PCLOSE ($\geq$ 0.1)	0.919	0.419	0.668	0.460

†p < 0.1; *p < 0.05; **p < 0.01; ***p < 0.001

Technology base and futurity

	Model 1	Model 2	Model 3	Model 4
Technology orientation	0.105			0.087
Research intensity		0.037		-0.040
Product focus			$0.144^{\dagger}$	0.137
R^2	0.011	0.001	0.021	0.028
χ^2	1.3	5.6	2.1	7.2
df	2	2	2	6
P ($\geq$ 0.1)	0.631	0.062	0.345	0.306
χ^2/df ($\leq$ 3.0)	0.532	2.786	1.066	1.195
RMSEA ($\leq$ 0.1)	0.000	0.096	0.018	0.032
CFI ($\geq$ 0.9)	1.000	0.981	0.999	0.995
NFI ($\geq$ 0.9)	0.994	0.972	0.989	0.974
PCLOSE ($\geq$ 0.1)	0.669	0.150	0.502	0.578

†p < 0.1; *p < 0.05; **p < 0.01; ***p < 0.001

Technology base and flexibility

	Model 1	Model 2	Model 3	Model 4
Technology orientation	0.133			0.207
Research intensity		-0.013		-0.003
Product focus			-0.088	-0.115
R^2	0.018	0.000	0.008	0.042
χ^2				0.9
df	0	0	0	2
P (≥ 0.1)				0.627
χ^2/df (≤ 3.0)				0.467
RMSEA (≤ 0.1)				0.000
CFI (≥ 0.9)				1.000
NFI (≥ 0.9)				0.994
PCLOSE (≥ 0.1)				0.744

$^\dagger p < 0.1$; $^* p < 0.05$; $^{**} p < 0.01$; $^{***} p < 0.001$

Technology base and partnering

	Model 1	Model 2	Model 3	Model 4
Technology orientation	0.252**			0.236*
Research intensity		0.120		0.011
Product focus			0.092	0.036
R^2	0.064	0.014	0.008	0.065
χ^2				0.1
df	0	0	0	2
P (≥ 0.1)				0.939
χ^2/df (≤ 3.0)				0.063
RMSEA (≤ 0.1)				0.000
CFI (≥ 0.9)				1.000
NFI (≥ 0.9)				0.999
PCLOSE (≥ 0.1)				0.961

$^\dagger p < 0.1$; $^* p < 0.05$; $^{**} p < 0.01$; $^{***} p < 0.001$

Questionnaire:

Research-based Spin-offs in Europe

Survey on Strategy, Development and Performance of
Research-based Spin-off Companies in Europe

Prof. Dr. Michael Schefczyk
Dipl.-Wirt.-Ing. Rigo Tietz

I. GENERAL INFORMATION

According to our pre-tests filling in this **survey needs only 10-15 minutes of your time**. You can easily fill in this PDF-formular electronically and press the "Send per email" button at the end of this questionnaire, so only the data will be submitted. You can also print out the entire document and send it via post or fax.

University of Technology Dresden
Chair for Entrepreneurship and Innovation
Dipl.-Wirt.-Ing. Rigo Tietz
01069 Dresden
Fax +49 351 463-36883
Email rigo.tietz@tu-dresden.de

We guarantee to **treat the disclosed information strictly confidential and anonymous.** This study is financed by research fundings and thus totally free from any kind of commercial interests, so the results will be used for research purposes only.

If you have any further questions, please contact us. Mr Rigo Tietz, a colleague in our research group, would be glad to answer all your questions via phone +49 351 463-36882 or email rigo.tietz@tu-dresden.de.

We are very grateful for your help. By participating at our survey you can **personally benefit in several ways.** Please indicate how we can thank you for your support.

☐ Send me the report on Academic Entrepreneurship in Europe (DeCleyn/Tietz, 2010, 96 pages, digital version is immediately available).

☐ Send me the final report on this pan-european study about research-based spin-off companies.

☐ Please add me to your mailing list and keep me informed about research activities in the field of Academic Entrepreneurship.

☐ Please provide an individual analysis of my company compared to the examined spin-off population in 24 European countries (in this case we need to know the name of your company and your email address).

Company name: [   ]

Email address: [   ]

TECHNISCHE
UNIVERSITÄT
DRESDEN

A.   INFORMATION ABOUT YOUR COMPANY

A1.   Academic parent institutions
Please indicate which type of institution your company spun off (multiple answers possible)

- ☐ University
- ☐ University for Applied Science
- ☐ Non-University Research Institution
- ☐ Others: _____

A2.   Facts about the founding of the spin-off
Please fill in the following figures

| Date of founding (month/year) | _____ | | Number of founders | _____ |

Number of external entrepreneurs
from outside the parent institutions
in the founding team   _____   →   External entrepreneur = Individual co-founder of the spin-off company who did not belong to the parent institutions before founding

Number of institutional investors
at founding   _____   →   Institutional investor = Non-individual co-founder of the spin-off company (e.g. Venture Capitalist, parent institution)

A3.   Characteristics of your products / services
Please indicate which statement suits mostly to your current portfolio

	1	2	3	4	5	6	7	
The current portfolio of our company is only based on services.	☐	☐	☐	☐	☐	☐	☐	The current portfolio of our company is only based on products.
Our portfolio consists of low technology products and services.	☐	☐	☐	☐	☐	☐	☐	Our portfolio consists of high technology products and services.
We developed our portfolio completely without research efforts.	☐	☐	☐	☐	☐	☐	☐	We developed our portfolio on the basis of extensive research efforts.

A4.   Main industry sector of your company
Please indicate which industry your company belongs to (multiple answers possible)

- ☐ Manufacturing
- ☐ Biotechnology / Pharmaceutical
- ☐ Microelectronic
- ☐ IT / Telecommunication
- ☐ Software
- ☐ Others: _____

A5.   Current life-stage of your company
Please choose the answer which suits mostly to the current situation of your company

■ Start-up	■ Expansion	■ Maturity	■ Diversification	■ Decline
Formal establishment, informal structure, focus on obtaining resources and building prototype, inconsistent growth rates	Functionally organised structure emerges, early formalisation of policies, focus on volume production, capacity expansion, rapid growth rate	More bureaucratic organisation, focus on making business profi-table, expense control, establishing manage-ment system, slow growth rate	Divisionalisation, diversification of products and market scope, use of sophisticated controls and planning system, high rowth rate	Demand for products levels off, low rate of innovation, profitability starts to drop, redefinition of mission and strategy, declining growth rate

A6.   Your current position in the company: _____

TECHNISCHE UNIVERSITÄT DRESDEN

B.   MANAGEMENT TEAM

B1.   Structure of the management team
Please consider today's situation

Size of the management team → Management team = All current managing directors (e.g. CEO, CFO, CTO, CSO, COO) of the spin-off company

Number of external managers from outside the parent institutions → External manager = Member of the management team who did not belong to the parent institution before joining the spin-off company

B2.   Characteristics of the management team
Please consider today's situation for each member of the management team

	Manager 1	Manager 2	Manager 3	Manager 4	Manager 5	Manager 6
Member of the original founding team	☐	☐	☐	☐	☐	☐
External manager (from outside the parent institutions)	☐	☐	☐	☐	☐	☐
(Partly) owner of the company	☐	☐	☐	☐	☐	☐
Current position in the company (e.g. CEO, CTO, COO)						
Number of years in the company						
Age (if known)						

B3.   Qualification of the management team
Please consider today's situation for each member of the management team

	Manager 1	Manager 2	Manager 3	Manager 4	Manager 5	Manager 6
Highest degree: BA = Bachelor, MA = Master/diploma, PhD = Doctor						
Degree in management / business / economics	☐	☐	☐	☐	☐	☐
Degree in science / engineering / informatics / medical	☐	☐	☐	☐	☐	☐

B4.   Previous experience of the management team
Please consider the experience of each member of the management team before joining the spin-off

	Manager 1	Manager 2	Manager 3	Manager 4	Manager 5	Manager 6
Previous entrepreneurial experience in launching an own start-up business	☐	☐	☐	☐	☐	☐
Previous experience in (partly) owning a business	☐	☐	☐	☐	☐	☐
Previous experience in successfully selling or merging the own business	☐	☐	☐	☐	☐	☐
Previous experience with a failure of the own business	☐	☐	☐	☐	☐	☐
Previous entrepreneurial experience in the same industry	☐	☐	☐	☐	☐	☐
Previous work experience in a management position	☐	☐	☐	☐	☐	☐
Previous industry work experience in a technical position	☐	☐	☐	☐	☐	☐

**TECHNISCHE
UNIVERSITÄT
DRESDEN**

C.   COMPANY STRATEGY

C1.   **Strategic orientation of your company**
Please choose a number to indicate whether you agree or disagree with the following statements. Marking (1) indicates a strong disagreement with the statement, whereas (7) indicates strong agreement, and (4) indicates that you neither agree nor disagree. The numbers in between, represent differing degrees of agreement.

| | 1 = Strongly disagree | | | | | 7 = Strongly agree | |
	1	2	3	4	5	6	7
Formal strategic plans serve as the basis for our competetive analysis.	⬜	⬜	⬜	⬜	⬜	⬜	⬜
We have a clear and consistent vision for where we want to end up with our business.	⬜	⬜	⬜	⬜	⬜	⬜	⬜
Our business strategy is typically not planned in advance, but emerges over time.	⬜	⬜	⬜	⬜	⬜	⬜	⬜
When confronted with a major decision, we usually try to develop it through analysis.	⬜	⬜	⬜	⬜	⬜	⬜	⬜
We organise and implement control processes to make sure that we meet our objectives.	⬜	⬜	⬜	⬜	⬜	⬜	⬜
Our business strategy is carefully planned before significant competetive actions are taken.	⬜	⬜	⬜	⬜	⬜	⬜	⬜

| | 1 = Strongly disagree | | | | | 7 = Strongly agree | |
	1	2	3	4	5	6	7
Our criteria for resource allocation generally reflect short-term considerations.	⬜	⬜	⬜	⬜	⬜	⬜	⬜
Forecasting key indicators of operations is common in our company.	⬜	⬜	⬜	⬜	⬜	⬜	⬜
Formal tracking of significant general trends is common in our company.	⬜	⬜	⬜	⬜	⬜	⬜	⬜
We often conduct "what if" analyses of critical issues.	⬜	⬜	⬜	⬜	⬜	⬜	⬜
We use pre-commitments from customers and suppliers as often as possible.	⬜	⬜	⬜	⬜	⬜	⬜	⬜
We use agreements with customers, suppliers, and others to reduce the amount of uncertainty.	⬜	⬜	⬜	⬜	⬜	⬜	⬜

| | 1 = Strongly disagree | | | | | 7 = Strongly agree | |
	1	2	3	4	5	6	7
We experiment with different products and/or services.	⬜	⬜	⬜	⬜	⬜	⬜	⬜
The products / services we now provide are substantially different than we first imagined.	⬜	⬜	⬜	⬜	⬜	⬜	⬜
We allow the business to evolve as opportunities emerge.	⬜	⬜	⬜	⬜	⬜	⬜	⬜
We adapt what we are doing to the resources we have.	⬜	⬜	⬜	⬜	⬜	⬜	⬜
We are flexible and take advantage of opportunities as they arise.	⬜	⬜	⬜	⬜	⬜	⬜	⬜
We avoid courses of action that restrict our flexibility and adaptability.	⬜	⬜	⬜	⬜	⬜	⬜	⬜

TECHNISCHE
UNIVERSITÄT
DRESDEN

C2.   Strategic posture of your company
Please choose a number to indicate which of the two statements is most true for your company.
Marking (1) indicates a strong agreement with the first statement, whereas (7) indicates strong
agreement with the second statement, and (4) indicates both are equally true. The numbers in
between, represent differing degrees of agreement with one of the two statements.

First statement more true	1	2	3	4	5	6	7	Second statement more true
Typically, we respond to actions which competitors initiate.	□	□	□	□	□	□	□	Typically, we initiate actions which competitors then respond to.
Our company is very seldom the first one to introduce new products/services, operating technologies, etc.	□	□	□	□	□	□	□	Our company is very often the first one to introduce new products/services, operating technologies, etc.
Our company has a strong tendency to "follow the leader" in introducing new products/services or ideas.	□	□	□	□	□	□	□	Our company has a strong tendency to be ahead of other competitors in introducing novel ideas or products/services
Generally, our company prefers to strongly emphasize the marketing of tried and true products or services.	□	□	□	□	□	□	□	Generally, our company prefers to strongly emphasize R&D, technological leadership, and innovations.
In the last 3 years, our company has introduced no new lines of products or services.	□	□	□	□	□	□	□	In the last 3 years, our company has introduced very many new lines of products or services.
Changes in products or services have been mostly of a minor nature.	□	□	□	□	□	□	□	Changes in products or services have usually been quite dramatic.

	1	2	3	4	5	6	7	
Normally, we seek to avoid competetive clashes, preferring a "live-and-let-live" posture.	□	□	□	□	□	□	□	Normally, we adopt a very competitive, "beat-the-competitor" posture.
Our actions towards other competitors can be termed aggressive.	□	□	□	□	□	□	□	Our actions towards other competitors can be termed accomodating.
Generally, our company has a strong tendency towards projects with low risk (with normal and certain results).	□	□	□	□	□	□	□	Generally, our company has a strong tendency towards high risk projects (with a chance for high results).
We generally believe, that it is best to achieve our objectives gradually via cautios, incremental behavior.	□	□	□	□	□	□	□	We generally believe, that it is best to achieve our objectives via bold and wide-ranging actions.
We typically adopt a fearless, aggressive position in order to maximise the probability of exploiting opportunities.	□	□	□	□	□	□	□	We typically adopt a "wait-and-see" position in order to minimise the probability of making costly decisions.
In our company, very many changes suggested by our employees are implemented.	□	□	□	□	□	□	□	In our company, very few changes suggested by our employees are implemented.
Identifying new products and services is the responsibility of a small number of individuals.	□	□	□	□	□	□	□	Identifying new products and services is done by all members of the company.
Our company discourages independent activity to develop new services and products.	□	□	□	□	□	□	□	Our company encourages independent activity to develop new services and products.

TECHNISCHE
UNIVERSITÄT
DRESDEN

D.   COMPANY DEVELOPMENT AND PERFORMANCE

D1.   **Environmental situation** for your company
Please indicate which of the following statements is most true for your company

	1	2	3	4	5	6	7	
The demand for our products/services is steady over time.	☐	☐	☐	☐	☐	☐	☐	The rate of demand for our products/ services fluctuates significantly.
We can predict our customers' preferences and demands with regard to our products/services over time.	☐	☐	☐	☐	☐	☐	☐	It is not possible to predict in advance demand changes regarding our products/services.
Future technological innovations affecting the viability of our products/services are likely to be frequent and major.	☐	☐	☐	☐	☐	☐	☐	Future technological innovations affecting the viability of our products/services are likely to be rare and minor.
It is not possible to predict kinds or timing of future innovations that will affect the viability of our products/services.	☐	☐	☐	☐	☐	☐	☐	We can predict the nature and source of innovations that affect the viability of our products/services.

D2.   **Performance of your company**
Please compare the performance of your company to the performance of your competitors

Compared to our Competitors...	1 = Much worse			4 = Equal			7 = Much better
	1	2	3	4	5	6	7
...Growth in Sales is...	☐	☐	☐	☐	☐	☐	☐
...Growth in Employees is...	☐	☐	☐	☐	☐	☐	☐
...Growth in Market Share is...	☐	☐	☐	☐	☐	☐	☐
...Net Profits are...	☐	☐	☐	☐	☐	☐	☐
...Return on Investment is...	☐	☐	☐	☐	☐	☐	☐

D3.   **Figures about your company**
If possibile please provide the following figures of your company

Year	Total turnover (EUR)	Total number of employees (FTE)
2009		
2008		
2007		
2006		
Within 12 months after founding		

FTE = Full-time equivalent
(including management team)

II.   FINAL COMMENTS

Thanks again for your time and effort! In order to submit the data you can either press the "Send per email" button or print out the document and send it via post or fax.

Send data p er email   **Print document**

University of Technology Dresden
Chair for Entrepreneurship and Innovation
Rigo Tietz
01069 Dresden
Fax 0049 351 463-36883
Tel. 0049 351 463-36882
Email: rigo.tietz@tu-dresden.de

References

Aaboen, L., Lindelöf, P., von Koch, C., and Löfsten, H. (2006). Corporate governance and performance of small high-tech firms in Sweden. *Technovation, 26*(8), 955-968.

Ackelsberg, R., and Arlow, P. (1985). Small businesses do plan and it pays off. *Long Range Planning, 18*(5), 61-67.

Acs, Z. J., and Audretsch, D. B. (1989). Entrepreneurial strategy and the presence of small firms. *Small Business Economics, 1*(3), 193-213.

Adizes, I. (1989). *Corporate lifecycles: How and why corporations grow and die and what to do about it.* Englewood Cliffs, NY

Agrawal, A. (2001). University-to-industry knowledge transfer: Literature review and unanswered questions. *International Journal of Management Reviews, 3*(4), 285-302.

Aldrich, H., and Auster, E. R. (1986). Even dwarfs started small: Liabilities of age and size and their strategic implications. *Research in Organizational Behavior, 8*, 165-198.

Allison, G. T. (1971). *Essence of decision: Explaining the Cuba missile crisis.* Boston: Little, Brown.

Almus, M., and Nerlinger, E. A. (1999). Growth of new technology-based firms: Which factors matter? *Small Business Economics, 13*(2), 141-154.

Amason, A. C., Shrader, R. C., and Tompson, G. H. (2006). Newness and novelty: Relating top management team composition to new venture performance. *Journal of Business Venturing, 21*(1), 125-148.

Analoui, F., and Karami, A. (2003). *Strategic management in small and medium enterprises.* London: Thomson.

Andersen, T. J., and Nielsen, B. B. (2009). Adaptive strategy making: The effects of emergent and intended strategy modes. *European Management Review, 6*(2), 94-106.

Andersen, T. J. (2004). Integrating decentralized strategy making and strategic planning processes in dynamic environments. *Journal of Management Studies, 41*(8), 1271-1299.

Andrews, K. R. (1971). *The concept of corporate strategy.* Homewood, IL: Irwin.

Ansoff, H. I. (1994). Comment on Henry Mintzberg's rethinking strategic planning. *Long Range Planning, 27*(3), 31-32.

Ansoff, H. I. (1965). *Corporate strategy - an analytical approach to business policy for growth and expansion*. New York: McGraw-Hill.

Ansoff, H. I. (1991). Critique of Henry Mintzberg's 'the design school: Reconsidering the basic premises of strategic management'. *Strategic Management Journal, 12*(6), 449-461.

Aragón-Sánchez, A., and Sánchez-Marín, G. (2005). Strategic orientation, management characteristics, and performance: A study of Spanish SMEs. *Journal of Small Business Management, 43*(3), 287-308.

Arbuckle, J. L. (2009). *Amos 18 User's guide*. Chicago, IL: SPSS Inc.

Armstrong, J. S. (1982). The value of formal planning for strategic decisions: Review of empirical research. *Strategic Management Journal, 3*(3), 197-211.

Armstrong, J. S., and Overton, T. S. (1977). Estimating nonresponse bias in mail surveys. *Journal of Marketing Research, 14*(3), 396-402.

Aspelund, A., Berg-Utby, T., and Skjevdal, R. (2005). Initial resources' influence on new venture survival: A longitudinal study of new technology-based firms. *Technovation, 25*(11), 1337-1347.

Astley, W. G. (1984). Toward an appreciation of collective strategy. *Academy of Management Review, 9*(3), 526-535.

Audretsch, D. B. (2000). *Is university entrepreneurship different?* Mimeograph, Indiana University.

Autio, E. (1994). New, technology-based firms as agents of R&D and innovation: An empirical study. *Technovation, 14*(4), 259-273.

Autio, E., and Lumme, A. (1998). Does the innovator role affect the perceived potential for growth? Analysis of four types of new technology-based firms. *Technology Analysis & Strategic Management, 10*(1), 41.

Backhaus, K., Blechschmidt, B., and Eisenbeiß, M. (2006). Der Stichprobeneinfluss bei Kausalanalysen. *Die Betriebswirtschaft, 66*(6), 711-726.

Backhaus, K., Erichson, B., Plinke, W., and Weiber, R. (2003). *Multivariate Analysemethoden: Eine anwendungsorientierte Einführung*. Berlin: Springer.

Bagozzi, R. P., and Baumgartner, H. (1994). The evaluation of structural equation models and hypothesis testing. In R. P. Bagozzi (Ed.), *Principles in marketing research* (pp. 386-422). Cambridge.

Bagozzi, R. P., and Yi, Y. (1988). On the evaluation of structural equation models. *Academy of Marketing Science.Journal, 16*(1), 74-94.

Baker, W. H., Addams, H. L., and Davis, B. (1993). Business planning in successful small firms. *Long Range Planning, 26*(6), 82-88.

Balderjahn, I. (1986). *Das umweltbewusste Konsumentenverhalten.* Berlin.

Bantel, K. A., and Jackson, S. E. (1989). Top management and innovations in banking: Does the composition of the top team make a difference? *Strategic Management Journal, 10*, 107-124.

Barker III, V. L., and Mueller, G. C. (2002). CEO characteristics and firm R&D spending. *Management Science, 48*(6), 782-801.

Barnard, C. (1938). *The functions of the executives.* Harvard University Press.

Barney, J. B. (1986). Strategic factor markets: Expectations, luck, and business strategy. *Management Science, 32*(10), 1231-1241.

Barney, J. B. (1991). Firm resources and sustained competitive advantage. *Journal of Management, 17*(1), 99-120.

Barney, J. B. (2001a). Is the resource-based 'view' a useful perspective for strategic management research? Yes. *Academy of Management Review, 26*(1), 41-56.

Barney, J. B. (2001b). Resource-based theories of competitive advantage: A ten-year retrospective on the resource-based view. *Journal of Management, 27*(6), 643-650.

Baron, R. A., and Ensley, M. D. (2006). Opportunity recognition as the detection of meaningful patterns: Evidence from comparisons of novice and experienced entrepreneurs. *Management Science, 52*(9), 1331-1344.

Barringer, B. R., Jones, F. F., and Neubaum, D. O. (2005). A quantitative content analysis of the characteristics of rapid-growth firms and their founders. *Journal of Business Venturing, 20*(5), 663-687.

Bates, T. (1990). Entrepreneur human capital inputs and small business longevity. *Review of Economics & Statistics, 72*(4), 551-559.

Batjargal, B. (2007). Internet entrepreneurship: Social capital, human capital, and performance of internet ventures in china. *Research Policy, 36*(5), 605-618.

Bauer, H. H., and Wölfer, H. (2001). Möglichkeiten und Grenzen der Online-Marktforschung. *Management Know-how, Nr. M 58.* Universität Mannheim: Institut für Marktorientierte Unternehmensführung.

Bearden, W. O., Netemeyer, R. G., and Teel, J. E. (1989). Measurement of consumer susceptibility to interpersonal influence. *Journal of Consumer Research, 15*(4), 473-481.

Becker, G. S. (1964). *Human capital.* New York: Columbia University Press.

Becker, G. S. (1975). *Human capital.* New York: National Bureau of Economic Research.

Beckman, C. M., Burton, M. D., and O'Reilly, C. (2007). Early teams: The impact of team demography on VC financing and going public. *Journal of Business Venturing, 22*(2), 147-173.

Begley, T. M., and Boyd, D. P. (1987). Psychological characteristics associated with performance in entrepreneurial firms and smaller businesses. *Journal of Business Venturing, 2*(1), 79-93.

Bellini, E., Capaldo, G., Edström, A., Kaulio, M., Raffa, M., Riccardia, M., et al. (1999). Strategic paths of academic spin-offs: A comparative analysis of Italian and Swedish cases. *44th ICSB Conference, Naples, 20-23 June.*

Bentler, P. M., and Bonett, D. G. (1980). Significance tests and goodness of fit in the analysis of covariance structures. *Psychological Bulletin, 88*(3), 588-606.

Berndts, P., and Harmsen, D. M. (1985). *Technologieorientierte Unternehmensgründungen in Zusammenarbeit mit staatlichen Forschungseinrichtungen.* Köln.

Berry, M. (1998). Strategic planning in small high tech companies. *Long Range Planning, 31*(3), 455-466.

Bertrand, M., and Schoar, A. (2003). Managing with style: The effect of managers on firm policies. *Quarterly Journal of Economics, 118*(4), 1169-1208.

Bhide, A. V. (2000). *The origin and evolution of new businesses.* New York, NY: Oxford University Press.

Biggadike, E. R. (1979). *Corporate diversification: Entry, strategy, and performance.* Cambridge.

Blair, D. M., and Hitchens, D. M. (1998). *Campus companies - UK and Ireland.* Ashgate, Aldershot.

Blau, P. (1977). *Inequality and heterogeneity.* New York: Free Press.

Boeker, W., and Karichalil, R. (2002). Entrepreneurial transitions: Factors influencing founder departure. *Academy of Management Journal, 45*(4), 818-826.

Boeker, W., and Wiltbank, R. (2005). New venture evolution and managerial capabilities. *Organization Science, 16*(2), 123-133.

Bollinger, L., Hope, K., and Utterback, J. M. (1983). A review of literature and hypotheses on new technology-based firms. *Research Policy, 12*(1), 1-14.

Borch, O. J., Huse, M., and Senneseth, K. (1999). Resource configuration, competitive strategies, and corporate entrepreneurship: An empirical examination of small firms. *Entrepreneurship: Theory & Practice, 24*(1), 51-72.

Bortz, J., and Döring, N. (2006). *Forschungsmethoden und Evaluation für Human- und Sozialwissenschaftler.* Heidelberg: Springer Verlag.

Boulton, W. R., Lindsay, W. M., Franklin, S. G., and Rue, L. W. (1982). Strategic planning: Determining the impact of environmental characteristics and uncertainty. *Academy of Management Journal, 25*(3), 500-509.

Bower, J. L. (1970). *Managing the resource allocation process.* Boston: Havard Business School Press.

Bower, D. J. (2003). Business model fashion and the academic spinout firm. *R&D Management, 33*(2), 97-106.

Boyd, B. K. (1991). Strategic planning and financial performance: A meta-analytic review. *Journal of Management Studies, 28*(4), 353-374.

Boyd, B. K., Dess, G. G., and Rasheed, A. M. A. (1993). Divergence between archival and perceptual measures of the environment: Causes and consequences. *Academy of Management Review, 18*(2), 204-226.

Boyd, B. K., Gove, S., and Hitt, M. A. (2005). Consequences of measurement problems in strategic management research: The case of Amihud and Lev. *Strategic Management Journal, 26*(4), 367-375.

Bracker, J. S., Keats, B. W., and Pearson, J. N. (1988). Planning and financial performance among small firms in a growth industry. *Strategic Management Journal, 9*(6), 591-603.

Bracker, J. S., and Pearson, J. N. (1986). Planning and financial performance of small, mature firms. *Strategic Management Journal, 7*(6), 503-522.

Bray, M. J., and Lee, J. N. (2000). University revenues from technology transfer: Licensing fees vs. equity positions. *Journal of Business Venturing, 15*(5-6), 385-392.

Brews, P. J., and Hunt, M. R. (1999). Learning to plan and planning to learn: Resolving the planning school/learning school debate. *Strategic Management Journal, 20*(10), 889-913.

Brinckmann, J., Grichnik, D., and Kapsa, D. (2010). Should entrepreneurs plan or just storm the castle? A meta-analysis on contextual factors impacting the business planning-performance relationship in small firms. *Journal of Business Venturing, 25*(1), 24-40.

Brouthers, K. D., Andriessen, F., and Nicolaes, I. (1998). Driving blind: Strategic decision making in small companies. *Long Range Planning, 31*(1), 130-138.

Brouthers, K. D., Brouthers, L. E., and Werner, S. (2000). Influences on strategic decision-making in the Dutch financial services industry. *Journal of Management, 26*(5), 863-883.

Brown, T. E., Davidsson, P., and Wiklund, J. (2001). An operationalization of Stevenson's conceptualization of entrepreneurship as opportunity-based firm behavior. *Strategic Management Journal, 22*(10), 953-968.

Browne, M. W., and Cudeck, R. (1993). Alternative ways of assessing equation model fit. In K. Bollen, and J. S. Long (Eds.), *Testing structural equation models* (pp. 136-162). Newbury Park.

Brüderl, J., and Preisendörfer, P. (2000). Fast-growing businesses. *International Journal of Sociology, 30*, 45-70.

Brüderl, J., Preisendörfer, P., and Ziegler, R. (1992). Survival chances of newly funded business organizations. *American Sociological Review, (57)*, 227-242.

Brüderl, J., Preisendörfer, P., and Ziegler, R. (1996). *Der Erfolg neugegründeter Betriebe: Eine empirische Studie zu den Chancen und Risiken von Unternehmensgründungen.* Berlin.

Brüderl, J., and Schüssler, R. (1990). Organizational mortality: The liabilities of newness and adolescence. *Administrative Science Quarterly, 35*(3), 530-547.

Bruno, A. V., Leidecker, J. K., and Harder, J. W. (1987). Why firms fail. *Business Horizons, 30*(2), 50-58.

Brush, C. G., Greene, P. G., and Hart, M. M. (2001). From initial idea to unique advantage: The entrepreneurial challenge of constructing a resource base. *Academy of Management Executive, 15*(1), 64-78.

Bruton, G. D., and Rubanik, Y. (2002). Resources of the firm, Russian high-technology startups, and firm growth. *Journal of Business Venturing, 17*(6), 553-576.

Burgelman, R. A. (1983). A process model of internal corporate venturing in the diversified major firm. *Administrative Science Quarterly, 28*(2), 223-244.

Campbell, A., and Faulkner, D. O. (2003). Introduction to Vol. 1: Competitive strategy through different lenses. In D. O. Faulkner, and A. Campbell (Eds.), *The Oxford handbook of strategy, Vol. 1* (pp. 1-20). Oxford.

Cannella, A. A., and Holcomb, T. R. (2005). A multi-level analysis of the upper echelon model. In F. Dansereau, and F. Yammarino (Eds.), *Multi-level issues in strategy and methods*. Oxford: Elsevier.

Cannella, B., Finkelstein, S., and Hambrick, D. C. (2008). *Strategic leadership*. Oxford: Oxford University Press.

Cantillon, R. (1755). *Essai sur la nature du commerce en general*.

Carayannis, E. G., Rogers, E. M., Kurihara, K., and Allbritton, M. M. (1998). High-technology spin-offs from government R&D laboratories and research universities. *Technovation, 18*(1), 1-11.

Carlsson, G., and Karlsson, K. (1970). Age, cohorts and the generation of generations. *American Sociological Review, 35*(4), pp. 710-718.

Carpenter, M. A., Geletkancz, M. A., and Sanders, W. G. (2004). Upper echelons research revisited: Antecedents, elements, and consequences of top management team composition. *Journal of Management, 30*(6), 749-778.

Carpenter, M. A., and Reilly, G. P. (2006). Constructs and construct measurement in upper echelon research. In D. J. Ketchen, and D. D. Berg (Eds.), *Research methodology in strategy and management*. Oxford: Elsevier.

Chaffee, E. E. (1985). Three models of strategy. *Academy of Management Review, 10*(1), 89-98.

Chandler, A. D. (1962). *Strategy and structure*. Cambridge: The MIT Press.

Chandler, A. D. (1977). *The visible hand*. Cambridge: Harvard University Press.

Chandler, G. N., DeTienne, D. R., McKelvie, A., and Mumford, T. V. (2011). Causation and effectuation processes: A validation study. *Journal of Business Venturing, 26*(3), 375-390.

Chandler, G. N., and Hanks, S. H. (1993). Measuring the performance of emerging businesses: A validation study. *Journal of Business Venturing, 8*(5), 391-408.

Chandler, G. N., and Hanks, S. H. (1994). Founder competence, the environment, and venture performance. *Entrepreneurship: Theory & Practice, 18*(3), 77-89.

Chandler, G. N., and Hanks, S. H. (1998). An examination of the substitutability of founders human and financial capital in emerging business ventures. *Journal of Business Venturing, 13*(5), 353-369.

Chandler, G. N., Honig, B., and Wiklund, J. (2005). Antecedents, moderators, and performance consequences of membership change in new venture teams. *Journal of Business Venturing, 20*(5), 705-725.

Chandler, G. N., and Jansen, E. (1992). The founder's self-assessed competence and venture performance. *Journal of Business Venturing, 7*(3), 223-236.

Chiesa, V., and Piccaluga, A. (2000). Exploitation and diffusion of public research: The case of academic spin-off companies in Italy. *R&D Management, 30*(4), 329.

Child, J. (1972). Organizational structure, environment and performance: The role of strategic choice. *Sociology, 6*(1), 1-22.

Chowdhury, S. (2005). Demographic diversity for building an effective entrepreneurial team: Is it important? *Journal of Business Venturing, 20*(6), 727-746.

Chrisman, J. J., Hynes, T., and Fraser, S. (1995). Faculty entrepreneurship and economic development: The case of the University of Calgary. *Journal of Business Venturing, 10*(4), 267-281.

Clarysse, B., Heirman, A., and Degroof, J. (2000). An institutional and resource based explanation of growth patterns of research based spin-offs in Europe. In P. D. Reynolds, et al. (Eds.), *Frontiers of entrepreneurship research* (pp. 545-559). Babson College, MA: Center for Entrepreneurial Studies, Babson College.

Clarysse, B., and Moray, N. (2004). A process study of entrepreneurial team formation: The case of a research-based spin-off. *Journal of Business Venturing, 19*(1), 55-79.

Clarysse, B., Wright, M., Lockett, A., Van de Velde, E., and Vohora, A. (2005). Spinning out new ventures: A typology of incubation strategies from European research institutions. *Journal of Business Venturing, 20*(2), 183-216.

Collis, D. J. (1994). Research note: How valuable are organizational capabilities? *Strategic Management Journal, 15*(8), 143-152.

Colombo, M. G., and Delmastro, M. (2002). How effective are technology incubators?: Evidence from Italy. *Research Policy, 31*(7), 1103-1122.

Colombo, M. G., Delmastro, M., and Grilli, L. (2004). Entrepreneurs' human capital and the start-up size of new technology-based firms. *International Journal of Industrial Organization, 22*(8-9), 1183-1211.

Colombo, M. G., and Grilli, L. (2005). Founders' human capital and the growth of new technology-based firms: A competence-based view. *Research Policy, 34*(6), 795-816.

Conner, K. R. (1991). A historical comparison of resource-based theory and five schools of thought within industrial organization economics: Do we have a new theory of the firm? *Journal of Management, 17*(1), 121-154.

Cooper, A. C., and Bruno, A. V. (1977). Success among high-technology firms. *Business Horizons, 20*(2), 16-22.

Cooper, A. C., and Daily, C. M. (1997). Entrepreneurial teams. In S. D.L. & S. R.W. (Ed.), *Entrepreneurship 2000* (pp. 127-150). Chicago, Illinois: Upstart Publishing Company.

Cooper, A. C., Gimeno-Gascon, F. J., and Woo, C. Y. (1994). Initial human and financial capital as predictors of new venture performance. *Journal of Business Venturing, 9*(5), 371-395.

Courtney, H., Kirkland, J., and Viguerie, P. (1997). Strategy under uncertainty. *Harvard Business Review, 75*(6), 67-79.

Covin, J. G., and Covin, T. J. (1990). Competitive aggressiveness, environmental context, and small firm performance. *Entrepreneurship: Theory & Practice, 14*(4), 35-50.

Covin, J. G., and Lumpkin, G. T. (2011). Entrepreneurial orientation theory and research: Reflections on a needed construct. *Entrepreneurship: Theory & Practice, 35*(5), 855-872.

Covin, J. G., and Slevin, D. P. (1989). Strategic management of small firms in hostile and benign environments. *Strategic Management Journal, 10*(1), 75-87.

Covin, J. G., and Slevin, D. P. (1990). New venture strategic posture, structure, and performance: An industry life cycle analysis. *Journal of Business Venturing, 5*(2), 123-135.

Covin, J. G., and Slevin, D. P. (1991). A conceptual model of entrepreneurship as firm behavior. *Entrepreneurship: Theory & Practice, 16*(1), 7-25.

Covin, J. G., Slevin, D. P., and Schultz, R. L. (1994). Implementing strategic missions: Effective strategic, structural and tactical choices. *Journal of Management Studies, 31*(4), 481-505.

Cressy, R. (1996). Are business startups debt-rationed. *Economic Journal, 106*(438), 1253-1270.

Cyert, R. M., and March, J. A. (1963). *A behavioural theory of the firm.* New York: Prentice-Hall: Englewood Cliffs.

Dahlstrand, Å. L. (1997). Growth and inventiveness in technology-based spin-off firms. *Research Policy, 26*(3), 331-344.

Daily, C. M., and Dalton, D. R. (1992). Financial performance of founder-managed versus professionally managed small corporations. *Journal of Small Business Management, 30*(2), 25-34.

Damanpour, F. (1991). Organizational innovation: A meta-analysis of effects of determinants and moderators. *Academy of Management Journal, 34*(3), 555-590.

Davidson, H. (1987). *Offensive marketing how to make your competitors followers.* London: Penguin.

Davidsson, P., and Honig, B. (2003). The role of social and human capital among nascent entrepreneurs. *Journal of Business Venturing, 18*(3), 301-331.

Davidsson, P., and Wiklund, J. (2001). Levels of analysis in entrepreneurship research: Current research practice and suggestions for the future. *Entrepreneurship: Theory & Practice, 25*(4), 81.

Davig, W. (1986). Business strategies in smaller manufacturing firms. *Journal of Small Business Management, 24*(1), 38-46.

De Cleyn, S. H. (2011). *The early development of academic spin-offs: A holistic study on the survival of 185 European product-oriented ventures using a resource-based perspective.* Universiteit Antwerpen.

De Cleyn, S. H., and Braet, J. (2009). Research valorisation through spin-off ventures: Integration of existing concepts and typologies. *World Review of Entrepreneurship, Management and Sust. Development, 5*(4), 325-352.

De Coster, R., and Butler, C. (2005). Assessment of proposals for new technology ventures in the UK: Characteristics of university spin-off companies. *Technovation, 25*(5), 535-543.

Dean Jr., J. W., and Sharfman, M. P. (1996). Does decision process matter? A study of strategic decision-making effectiveness. *Academy of Management Journal, 39*(2), 368-396.

Dean, C. C. (1993). *Corporate entrepreneurship: Strategic and structural correlates and impact on the global presence of United States firms.* Denton, TX: University of North Texas.

Debackere, K. (2000). Managing academic R&D as a business at K.U. Leuven: Context, structure and process. *R&D Management, 30*(4), 323-328.

Degroof, J., and Roberts, E. B. (2004). Overcoming weak entrepreneurial infrastructures for academic spin-off ventures. *Journal of Technology Transfer, 29*(3-4), 327-352.

Deshpandé, R., Farley, J. U., and Webster Jr., F. E. (1993). Corporate culture customer orientation, and innovativeness in Japanese firms: A quadrad analysis. *Journal of Marketing, 57*(1), 23-37.

Dess, G. G., and Beard, D. W. (1984). Dimensions of organizational task environments. *Administrative Science Quarterly, 29*(1), 52-73.

Dess, G. G., and Lumpkin, G. T. (2005). The role of entrepreneurial orientation in stimulating effective corporate entrepreneurship. *Academy of Management Executive, 19*(1), 147-156.

Dess, G. G., Lumpkin, G. T., and Covin, J. G. (1997). Entrepreneurial strategy making and firm performance: Tests of contingency and configurational models. *Strategic Management Journal, 18*(9), 677-695.

Dess, G. G., and Robinson Jr., R. B. (1984). Measuring organizational performance in the absence of objective measures: The case of the privately-held firm and conglomerate business unit. *Strategic Management Journal, 5*(3), 265-273.

Dew, N., Read, S., Sarasvathy, S. D., and Wiltbank, R. (2009). Effectual versus predictive logics in entrepreneurial decision-making: Differences between experts and novices. *Journal of Business Venturing, 24*(4), 287-309.

Di Gregorio, D., and Shane, S. (2003). Why do some universities generate more start-ups than others? *Research Policy, 32*(2), 209-227.

Diekmann, A. (2007). *Empirische Sozialforschung: Grundlagen, Methoden, Anwendungen.* Hamburg: Rowohlt Taschenbuch Verlag.

Dierickx, I., and Cool, K. (1989). Asset stock accumulation and sustainability of competitive advantage. *Management Science, 35*(12), 1504-1511.

Dimov, D. P., and Shepherd, D. A. (2005). Human capital theory and venture capital firms: Exploring "home runs" and "strike outs". *Journal of Business Venturing, 20*(1), 1-21.

Djokovic, D., and Souitaris, V. (2008). Spinouts from academic institutions: A literature review with suggestions for further research. *Journal of Technology Transfer, 33*(3), 225-247.

Dommeyer, C. J., and Moriarty, E. (1999). Comparing two forms of an e-mail survey: Embedded vs. attached. *International Journal of Market Research, 42*(1), 39-50.

Doty, D. H., and Glick, W. H. (1994). Typologies as a unique form of theory building: Toward improved understanding and modelling. *Academy of Management Review, 19*(2), 230-251.

Doutriaux, J. (1987). Growth pattern of academic entrepreneurial firms. *Journal of Business Venturing, 2*(4), 285-297.

Druilhe, C., and Garnsey, E. (2004). Do academic spin-outs differ and does it matter? *Journal of Technology Transfer, 29*(3-4), 269-285.

Dubini, P. (1989). Which venture capital backed entrepreneurs have the best chances of succeeding? *Journal of Business Venturing, 4*(2), 123-132.

Dyke, L. S., Fischer, E. M., and Reuber, R. (1992). An inter-industry examination of the impact of owner experience on firm performance. *Journal of Small Business Management, 30*(4), 72-87.

Edelman, L. F., Brush, C. G., and Manolova, T. (2005). Co-alignment in the resource–performance relationship: Strategy as mediator. *Journal of Business Venturing, 20*(3), 359-383.

Egeln, J., Gottschalk, S., Rammer, C., and Spielkamp, A. (2002). Spinoff-Gründungen aus der öffentlichen Forschung in Deutschland. Mannheim.

Eisenhardt, K. M. (1989). Making fast strategic decisions in high-velocity environments. *Academy of Management Journal, 32*(3), 543-576.

Eisenhardt, K. M., and Schoonhoven, C. B. (1990). Organizational growth: Linking founding team, strategy, environment, and growth among U.S. semiconductor ventures, 1978-1988. *Administrative Science Quarterly, 35*(3), 504-529.

Eisenhardt, K. M., and Zbaracki, M. J. (1992). Strategic decision making. *Strategic Management Journal, 13*, 17-37.

Elbanna, S. (2006). Strategic decision-making: Process perspectives. *International Journal of Management Reviews, 8*(1), 1-20.

Elfring, T., and Hulsink, W. (2003). Networks in entrepreneurship: The case of high-technology firms. *Small Business Economics, 21*(4), 409-422.

Ensley, M. D., Carland, J. C., and Carland, J. W. (1998). The effects of entrepreneurial team skill heterogeneity and functional diversity on new venture performance. *Journal of Business and Entrepreneurship, 10*(1), 1-11.

Ensley, M. D., and Hmieleski, K. M. (2005). A comparative study of new venture top management team composition, dynamics and performance between university-based and independent start-ups. *Research Policy, 34*(7), 1091-1105.

Ensley, M. D., and Pearson, A. W. (2005). An exploratory comparison of the behavioral dynamics of top management teams in family and nonfamily new ventures: Cohesion, conflict, potency, and consensus. *Entrepreneurship: Theory & Practice, 29*(3), 267-284.

Ensley, M. D., Pearson, A. W., and Amason, A. C. (2002). Understanding the dynamics of new venture top management teams: Cohesion, conflict, and new venture performance. *Journal of Business Venturing, 17*(4), 365-386.

Ensley, M. D., Pearson, A., and Pearce, C. L. (2003). Top management team process, shared leadership, and new venture performance: A theoretical model and research agenda. *Human Resource Management Review, 13*(2), 329.

Escribá-Esteve, A., Sánchez-Peinado, L., and Sánchez-Peinado, E. (2009). The influence of top management teams in the strategic orientation and performance of small and medium-sized enterprises. *British Journal of Management, 20*(4), 581-597.

Etzkowitz, H., and Klofsten, M. (2005). The innovating region: Toward a theory of knowledge-based regional development. *R&D Management, 35*(3), 243-255.

Farjoun, M. (2002). Towards an organic perspective on strategy. *Strategic Management Journal, 23*(7), 561.

Feldman, J. M., and Klofsten, M. (2000). *Medium-sized firms and the limits to growth: A case study in the evolution of a spin-off firm.* Routledge.

Ferrier, W. J. (2001). Navigating the competitive landscape: The drivers and consequences of competitive aggressiveness. *Academy of Management Journal, 44*(4), 858-877.

Finkelstein, S., and Hambrick, D. C. (1990). Top-management-team tenure and organizational outcomes: The moderating role of managerial discretion. *Administrative Science Quarterly, 35*(3), 484-503.

Finkelstein, S., and Hambrick, D. C. (1996). *Strategic leadership: Top executives and their effects on organizations.* Minneapolis: West Publishing.

Florin, J., Lubatkin, M., and Schulze, W. (2003). A social capital model of high-growth ventures. *Academy of Management Journal, 46*(3), 374-384.

Fontes, M. (2005). The process of transformation of scientific and technological knowledge into economic value conducted by biotechnology spin-offs. *Technovation, 25*(4), 339-347.

Forbes, D. P. (2005). Managerial determinants of decision speed in new ventures. *Strategic Management Journal, 26*(4), 355-366.

Fornell, C., and Bookstein, F. L. (1982). Two structural equation models: LISREL and PLS applied to consumer exit-voice theory. *Journal of Marketing Research, 19*(4), 440-452.

Fornell, C., and Larcker, D. F. (1981). Evaluating structural equation models with unobservable variables and measurement error. *Journal of Marketing Research), 18*, 39-50.

Franklin, S. J., Wright, M., and Lockett, A. (2001). Academic and surrogate entrepreneurs in university spin-out companies. *Journal of Technology Transfer, 26*(1-2), 127-141.

Fredrickson, J. W. (1984). The comprehensiveness of strategic decision processes: Extension, observations, future directions. *Academy of Management Journal, 27*(3), 445-466.

Fredrickson, J. W. (1986). An exploratory approach to measuring perceptions of strategic decision process constructs. *Strategic Management Journal, 7*(5), 473-483.

Fredrickson, J. W., and Iaquinto, A. L. (1989). Inertia and creeping rationality in strategic decision processes. *Academy of Management Journal, 32*(3), 516-542.

Fredrickson, J. W., and Mitchell, T. R. (1984). Strategic decision processes: Comprehensiveness and performance in an industry with an unstable environment. *Academy of Management Journal, 27*(2), 399-423.

Freeman, J., Carroll, G. R., and Hannan, M. T. (1983). The liability of newness: Age dependence in organizational death rates. *American Sociological Review, 48*, 692-710.

Furrer, O., Thomas, H., and Goussevskaia, A. (2008). The structure and evolution of the strategic management field: A content analysis of 26 years of strategic management research. *International Journal of Management Reviews, 10*(1), 1-23.

Gabarro, J. J. (1985). When a new manager takes charge. *Harvard Business Review, 63*(3), 110-123.

Galbraith, J., and Vesper, K. (1982). The stages of growth. *Journal of Business Strategy, 3*(1), 70-79.

Garengo, P., Biazzo, S., and Bititci, U. S. (2005). Performance measurement systems in SMEs: A review for a research agenda. *International Journal of Management Reviews, 7*(1), 25-47.

Gartner, W. B. (1985). A conceptual framework for describing the phenomenon of new venture creation. *Academy of Management Review, 10*(4), 696-706.

Gartner, W. B., Shaver, K. G., Gatewood, E., and Katz, J. A. (1994). Finding the entrepreneur in entrepreneurship. *Entrepreneurship: Theory & Practice, 18*(3), 5-9.

Gartner, W. B., Starr, J. A., and Bhat, S. (1999). Predicting new venture survival: An analysis of "anatomy of a start-up." cases from inc. magazine. *Journal of Business Venturing, 14*(2), 215-232.

Gary, A. K. (1997). Cross-cultural reliability and validity of a scale to measure firm entrepreneurial orientation. *Journal of Business Venturing, 12*(3), 213-225.

George, G., Wood Jr, D. R., and Khan, R. (2001). Networking strategy of boards: Implications for small and medium-sized enterprises. *Entrepreneurship & Regional Development, 13*(3), 269-285.

Gimeno, J., Folta, T. B., Cooper, A. C., and Woo, C. Y. (1997). Survival of the fittest? entrepreneurial human capital and the persistence of underperforming firms. *Administrative Science Quarterly, 42*(4), 750-783.

Golden, B. R., and Zajac, E. J. (2001). When will boards influence strategy? INCLINATION x POWER = STRATEGIC CHANGE. *Strategic Management Journal, 22*(12), 1087-1111.

Goldfarb, B., and Henrekson, M. (2003). Bottom-up versus top-down policies towards the commercialization of university intellectual property. *Research Policy, 32*(4), 639-658.

Goll, I., and Rasheed, A. M. A. (1997). Rational decision-making and firm performance: The moderating role of environment. *Strategic Management Journal, 18*(7), 583-591.

Grant, R. M. (2003). Strategic planning in a turbulent environment: Evidence from the oil majors. *Strategic Management Journal, 24*(6), 491.

Greiner, L. E. (1972). Evolution and revolution as organizations grow. *Harvard Business Review, 50*(4), 37-46.

Grupp, H., Jungmittag, A., Schmoch, U., and Legler, H. (2000). *Hochtechnologie 2000: Neudefinition der Hochtechnologie für die Berichterstattung zur technologischen Leistungsfähigkeit Deutschlands, Gutachten für das Bundesforschungsministerium (bmb+f)*. Karlsruhe, Hannover: Fraunhofer ISI und NIW.

Gustaffson, V. (2006). *Entrepreneurial decision-making: Individuals, tasks and cognitions.* Northampton, MA and Cheltenham: Edward Elgar.

Guzzo, R. A., and Dickson, M. W. (1996). Teams in organizations: Recent research on performance and effectiveness. *Annual Review of Psychology, 47*(1), 307.

Hage, J. (1980). *Theories of organizations.* New York: Wiley.

Hambrick, D. C., and Crozier, L. M. (1985). Stumblers and stars in the management of rapid growth. *Journal of Business Venturing, 1*(1), 31-45.

Hambrick, D. C., and Mason, P. A. (1984). Upper echelons: The organization as a reflection of its top managers. *Academy of Management Review, 9*(2), 193-206.

Hamel, G., and Prahalad, C. K. (1991). Corporate imagination and expeditionary marketing. *Harvard Business Review, 69*(4), 81-92.

Hanks, S. H., Watson, C. J., Jansen, E., and Chandler, G. N. (1993). Tightening the life-cycle construct: A taxonomic study of growth stage configurations in high-technology organizations. *Entrepreneurship: Theory & Practice, 18*(2), 5-30.

Hannan, M. T., and Freeman, J. (1977). The population ecology of organizations. *American Journal of Sociology, 82*(5), pp. 929-964.

Harper, D. A. (2008). Towards a theory of entrepreneurial teams. *Journal of Business Venturing, 23*(6), 613-626.

Harrison, R. T., and Leitch, C. M. (2005). Growth dynamics in university spin-out companies: Entrepreneurial ventures or technology lifestyle businesses? In B. Clarysse, J. B. Roure and T. Schamp (Eds.), *The role of private equity in growing new ventures*. Cheltenham: Edward Elgar Publishing.

Hart, S. L. (1992). An integrative framework for strategy-making processes. *Academy of Management Review, 17*(2), 327-351.

Hatten, K. J., and Schendel, D. E. (1977). Heterogeneity within an industry: Firm conduct in the U.S. brewing industry, 1952-71. *Journal of Industrial Economics, 26*(2), 97-113.

Hayes, R. H. (1985). Strategic planning--forward in reverse? *Harvard Business Review, 63*(6), 111-119.

He, L. (2008). Do founders matter? A study of executive compensation, governance structure and firm performance. *Journal of Business Venturing, 23*(3), 257-279.

Heirman, A., and Clarysse, B. (2004). How and why do research-based start-ups differ at founding? A resource-based configurational perspective. *Journal of Technology Transfer, 29*(3-4), 247-268.

Heirman, A., and Clarysse, B. (2007). Which tangible and intangible assets matter for innovation speed in start-ups? *Journal of Product Innovation Management, 24*(4), 303-315.

Hellerstedt, K. (2009). *The composition of new venture teams: Its dynamics and consequences*. Jönköping International Business School.

Helm, R., and Mauroner, O. (2007). Success of research-based spin-offs. state-of-the-art and guidelines for future research. *Review of Managerial Science, 1*(3), 237-270.

Hemer, J., Walter, G., Berteit, H., and Göthner, M. (2005). *Erfolgsfaktoren für Unternehmensausgründungen aus der Wissenschaft.* Fraunhofer-Institut für System- und Innovationsforschung.

Henry, E. (1998). The norms of entrepreneurial science: Cognitive effects of the new university–industry linkages. *Research Policy, 27*(8), 823-833.

Hickson, D., Butler, R., Cray, D., Mallory, G., and Wilson, D. (1986). *Top decisions: Strategic decision making in organizations.* San Francisco, CA: Jossey-Bass.

Hiller, N. J., and Hambrick, D. C. (2005). Conceptualizing executive hubris: The role of (hyper-)core self-evaluations in strategic decision-making. *Strategic Management Journal, 26*(4), 297-319.

Hindle, K., and Yencken, J. (2004). Public research commercialisation, entrepreneurship and new technology based firms: An integrated model. *Technovation, 24*(10), 793-803.

Hinings, C. R., and Greenwood, R. (1988). *The dynamics of strategic change.* Oxford: Basil Blackwell.

Hitt, M. A., Biermant, L., Shimizu, K., and Kochhar, R. (2001). Direct and moderating effects of human capital on strategy and performance in professional service firms: A resource-based perspective. *Academy of Management Journal, 44*(1), 13-28.

Hitt, M. A., Ireland, R. D., Camp, S. M., and Sexton, D. L. (2001). Guest editors' introduction to the special issue. Strategic Entrepreneurship: Entrepreneurial strategies for wealth creation. *Strategic Management Journal, 22*(6), 479-491.

Hitt, M. A., and Tyler, B. B. (1991). Strategic decision models: Integrating different perspectives. *Strategic Management Journal, 12*(5), 327-351.

Hofer, C. W., and Sandberg, W. R. (1987). Improving new venture performance: Some guidelines for success. *American Journal of Small Business, 12*(1), 11-25.

Homburg, C. (2000). *Kundennähe von Industriegüterunternehmen: Konzeption, Erfolgsaus-wirkungen, determinanten.* Wiesbaden: Gabler.

Homburg, C., and Giering, A. (1996). Konzeptualisierung und Operationalisierung komplexer Konstrukte: Ein Leitfaden für die Marketingforschung. *Marketing-Zeitschrift für Forschung und Praxis, 18*(1), 5-24.

Homburg, C., Pflesser, C., and Klarmann, M. (2008). Strukturgleichungsmodelle mit latenten Variablen: Kausalanalyse. In A. Herrmann, C. Homburg and M. Klarmann (Eds.), *Handbuch Marktforschung* (pp. 547-578). Wiesbaden: Gabler Verlag.

Homburg, C., and Baumgartner, H. (1995). Beurteilung von Kausalmodellen – Bestandsaufnahme und Anwendungsempfehlungen. *Marketing ZfP, 17*(3), 162-176.

Hopkins, W. E., and Hopkins, S. A. (1997). Strategic planning-financial performance relationships in banks: A causal examination. *Strategic Management Journal, 18*(8), 635-652.

Hornsby, J. S., Kuratko, D. F., and Zahra, S. A. (2002). Middle managers' perception of the internal environment for corporate entrepreneurship: Assessing a measurement scale. *Journal of Business Venturing, 17*(3), 253-273.

Hough, J. R., and White, M. A. (2003). Environmental dynamism and strategic decision-making rationality: An examination at the decision level. *Strategic Management Journal, 24*(5), 481.

Hsu, D. H. (2007). Experienced entrepreneurial founders, organizational capital, and venture capital funding. *Research Policy, 36*(5), 722-741.

Hubbard, R., Vetter, D. E., and Eldon L. Little. (1998). Replication in strategic management: Scientific testing for validity, generalizability, and usefulness. *Strategic Management Journal, 19*(3), 243-254.

Huff, A. (1990). *Mapping strategic thought.* Chichester, England: Wiley.

Huff, A. S., and Reger, R. K. (1987). A review of strategic process research. *Journal of Management, 13*(2), 211.

Hunsdiek, D. (1987). *Unternehmensgründung als Folgeinnovation. Struktur, Hemmnisse und Erfolgsbedingungen der Gründung industrieller Unternehmen.* Stuttgart: CE Poeschel.

Hunt, S. D., Sparkman Jr., R. D., and Wilcox, J. B. (1982). The pretest in survey research: Issues and preliminary findings. *Journal of Marketing Research (JMR), 19*(2), 269-273.

Hurley, R. F., and Hult, G. T. (1998). Innovation, market orientation, and organizational learning: An integration and empirical examination. *Journal of Marketing, 62*(3), 42-54.

Hutzschenreuter, T., and Kleindienst, I. (2006). Strategy-process research: What have we learned and what is still to be explored. *Journal of Management, 32*(5), 673-720.

Iaquinto, A. L., and Fredrickson, J. W. (1997). Top management team agreement about the strategic decision process: A test of some of its determinants and consequences. *Strategic Management Journal, 18*(1), 63-75.

Ireland, R. D., Hitt, M. A., and Sirmon, D. G. (2003). A model of strategic entrepreneurship: The construct and its dimensions. *Journal of Management, 29*(6), 963.

Jackson, S. (1992). Consequences of group composition for the interpersonal dynamics of strategic issue processing. In P. Shrivastava, A. Huff and J. Dutton (Eds.), *Advances in strategic management* (pp. 345-382). Greenwich, CT: JAI Press.

Jackson, S. E., Brett, J. F., Sessa, V. I., Cooper, D. M., Julin, J. A., and Peyronnin, K. (1991). Some differences make a difference: Individual dissimilarity and group heterogeneity as correlates of recruitment, promotions, and turnover. *Journal of Applied Psychology, 76*(5), 675-689.

Jagersma, P. K., and van Gorp, D. M. (2003). Spin-out management: Theory and practice. *Business Horizons, 46*(2), 15.

Jemison, D. B. (1981). Organizational versus environmental sources of influence in strategic decision making. *Strategic Management Journal, 2*(1), 77-89.

Johnson, G. (1987). *Strategic change and the management process.* Oxford: Basil Blackwell.

Jones-Evans, D., and Klofsten, M. (1999). Creating a bridge between university and industry in small European countries: The role of the industrial liaison office. *R&D Management, 29*(1), 47.

Kakati, M. (2003). Success criteria in high-tech new ventures. *Technovation, 23*(5), 447-457.

Kamm, J. B., Shuman, J. C., Seeger, J. A., and Nurick, A. J. (1990). Entrepreneurial teams in new venture creation: A research agenda. *Entrepreneurship: Theory & Practice, 14*(4), 7-17.

Kassicieh, S. K., Radosevich, R., and Banbury, C. M. (1997). Using attitudinal, situational, and personal characteristics variables to predict future entrepreneurs from national laboratory inventors. *IEEE Transactions of Engineering Management, 44*(3), 248-257.

Kay, R., May-Strobl, E., and Maaß, F. (2001). *Neue Ergebnisse der Existenzgründungsforschung.* Wiesbaden: Deutscher Universitäts-Verlag.

Kazanjian, R. K. (1988). Relation of dominant problems to stages growth in technology-based new ventures. *Academy of Management Journal, 31*(2), 257-279.

Keck, S. L. (1997). Top management team structure: Differential effects by environmental context. *Organization Science, 8*(2), 143-156.

Keeley, R. H., and Roure, J. B. (1990). Management, strategy, and industry structure: As influences on the success of new firms: A structural model. *Management Science, 36*(10), 1256-1267.

Khandwalla, P. N. (1976). Some top management styles, their context and performance. *Organization and Administrative Sciences, 7*(4), 21-51.

Kim, W. C., and Mauborgne, R. (2009). How strategy shapes structure. *Harvard Business Review, 87*(9), 72-80.

Kirchhoff, B. A. (1977). Organization effectiveness measurement and policy research. *Academy of Management Review, 2*(3), 347-355.

Kline, R. B. (2005). *Principles and praxis of structural equation modelling*. New York: Guilford Press.

Klofsten, M., and Jones-Evans, D. (2000). Comparing academic entrepreneurship in Europe - the case of Sweden and Ireland. *Small Business Economics, 14*(4), 299.

Klofsten, M., and Jones-Evans, D. (1996). Stimulation of technology-based small firms - A case study of university-industry cooperation. *Technovation, 16*(4), 187-193.

Knockaert, M., Ucbasaran, D., Wright, M., & Clarysse, B. (2011). The relationship between knowledge transfer, top management team composition, and performance: The case of science-based entrepreneurial firms. *Entrepreneurship Theory and Practice, 35*(4), 777-803.

Kotey, B., and Meredith, G. G. (1997). Relationships among owner/ manager personal values, business strategies, and enterprise performance. *Journal of Small Business Management, 35*(2), 37-64.

Kraus, S., Harms, R., and Schwarz, E. J. (2006). Strategic planning in smaller enterprises - new empirical findings. *Management Research News, 29*(6), 334-344.

Kraus, S., and Kauranen, I. (2009). Strategic management and entrepreneurship: Friends or foes? *International Journal of Business Science and Applied Management, 4*(1), 37-50.

Kreiser, P. M., Marino, L. D., and Weaver, K. M. (2002). Assessing the psychometric properties of the entrepreneurial orientation scale: A multi-country analysis. *Entrepreneurship: Theory & Practice, 26*(4), 71.

Kriegesmann, B. (2000). Unternehmensgründungen aus der Wissenschaft. Eine empirische Analyse zu Stand, Entwicklungen und institutionellen Rahmenbedingungen in außeruniversitären Forschungseinrichtungen. *Zeitschrift Für Betriebswirtschaft, 70*(4), 397-414.

Kulicke, M. (1987). *Technologieorientierte unternehmen in der Bundesrepublik Deutschland - eine empirische Untersuchung der Strukturbildungs- und Wachstumsphase von Neugründungen*. Frankfurt am Main: Peter Lang.

Kuratko, D. F., Ireland, R. D., Covin, J. G., and Hornsby, J. S. (2005). A model of middle-level managers' entrepreneurial behavior. *Entrepreneurship: Theory & Practice, 29*(6), 699-716.

Lee, C., Lee, K., and Pennings, J. M. (2001). Internal capabilities, external networks, and performance: A study on technology-based ventures. *Strategic Management Journal, 22*(6), 615.

Leitch, C. M., and Harrison, R. T. (2005). Maximising the potential of university spin-outs: The development of second-order commercialisation activities. *R&D Management, 35*(3), 257-272.

Lerner, J. (2005). The university and the start-up: Lessons from the past two decades. *Journal of Technology Transfer, 30*(1-2), 49-56.

Lieberman, M. B., and Montgomery, D. B. (1998). First-mover (dis)advantages: Retrospective and link with the resource-based view. *Strategic Management Journal, 19*(12), 1111.

Lindelof, P., and Lofsten, H. (2004). Proximity as a resource base for competitive advantage: University-industry links for technology transfer. *Journal of Technology Transfer, 29*(3-4), 311-326.

Link, A. N., and Scott, J. T. (2005). Opening the ivory tower's door: An analysis of the determinants of the formation of U.S. university spin-off companies. *Research Policy, 34*(7), 1106-1112.

Lockett, A., Murray, G., and Wright, M. (2002). Do UK venture capitalists still have a bias against investment in new technology firms. *Research Policy, 31*(6), 1009-1030.

Lockett, A., Siegel, D., Wright, M., and Ensley, M. D. (2005). The creation of spin-off firms at public research institutions: Managerial and policy implications. *Research Policy, 34*(7), 981-993.

Lockett, A., Thompson, S., and Morgenstern, U. (2009). The development of the resource-based view of the firm: A critical appraisal. *International Journal of Management Reviews, 11*(1), 9-28.

Lockett, A., Wright, M., and Franklin, S. (2003). Technology transfer and universities' spin-out strategies. *Small Business Economics, 20*(2), 185.

Loet, L. (2000). The triple helix: An evolutionary model of innovations. *Research Policy, 29*(2), 243-255.

Lovas, B., and Ghoshal, S. (2000). Strategy as guided evolution. *Strategic Management Journal, 21*(9), 875.

Lumpkin, G. T., and Dess, G. G. (1995). Simplicity as a strategy-making process: The effects of stage of organizational development and environment on performance. *Academy of Management Journal, 38*(5), 1386-1407.

Lumpkin, G. T., and Dess, G. G. (1996). Clarifying the entrepreneurial orientation construct and linking it to performance. *Academy of Management Review, 21*(1), 135-172.

Lumpkin, G. T., and Dess, G. G. (2001). Linking two dimensions of entrepreneurial orienttation to firm performance: The moderating role of environment and industry life cycle. *Journal of Business Venturing, 16*(5), 429-451.

Lyles, M. A., and Thomas, H. (1988). Strategic problem formulation: Biases and assumptions embedded in alternative decision-making models. *Journal of Management Studies, 25*(2), 131-145.

Lyon, D. W., Lumpkin, G. T., and Dess, G. G. (2000). Enhancing entrepreneurial orientation research: Operationalizing and measuring a key strategic decision making process. *Journal of Management, 26*(5), 1055-1085.

MacCallum, R. C., and Browne, M. W. (1993). The use of causal indicators in covariance structure models: Some practical issues. *Psychological Bulletin, 114*, 533-541.

Macmillan, I. C., Siegel, R., and Narasimha, P. N. S. (1985). Criteria used by venture capitalists to evaluate new venture proposals. *Journal of Business Venturing, 1*(1), 119-128.

Mahoney, J. T., and Pandian, J. R. (1992). The resource-based view within the conversation of strategic management. *Strategic Management Journal, 13*(5), 363-380.

March, J. G. (1962). The business firm as a political coalition. *The Journal of Politics, 24*, 662-678.

March, J. G., and Simon, H. A. (1958). *Organizations.* New York: John Wiley and Sons.

McDougall, P., and Robinson Jr., R. B. (1990). New venture strategies: An empirical identification of eight 'archetypes' of competitive strategies for entry. *Strategic Management Journal, 11*(6), 447-467.

McGrath, R. G., and MacMillan, I. (2000). *The entrepreneurial mindset.* London: Harvard Business School Press.

McGrath, R. G., and MacMillan, I. C. (1995). Discovery-driven planning. *Harvard Business Review, 73*(4), 44-54.

McKelvie, A., Haynie, J. M., and Gustavsson, V. (2011). Unpacking the uncertainty construct: Implications for entrepreneurial action. *Journal of Business Venturing, 26*(3), 273-292.

McMullen, J. S., and Shepherd, D. A. (2006). Entrepreneurial action and the role of uncertainty in the theory of the entrepreneur. *Academy of Management Review, 31*(1), 132-152.

McQueen, D. H., and Wallmark, J. T. (1982). Spin-off companies from Chalmers University of Technology. *Technovation, 1*(4), 305-315.

Metzger, G., Niefert, M., and Licht, G. (2008). *High-tech-Gründungen in Deutschand: Trends, Strukturen, Potenziale.* Mannheim: Zentrum für Europäische Wirtschafts-forschung (ZEW) GmbH.

Meyer, G. D., and Heppard, K. A. (2000). Entrepreneurial strategies - the dominant logic of entrepreneurship. In G. D. Meyer, and K. A. Heppard (Eds.), *Entrepreneurship as strategy - competing on the entrepreneurial edge* (pp. 1-22). London: Sage.

Meyer, M. (2003). Academic entrepreneurs or entrepreneurial academics? Research–based ventures and public support mechanisms. *R&D Management, 33*(2), 107-115.

Michel, J. G., and Hambrick, D. C. (1992). Diversification, posture and top management team characteristics. *Academy of Management Journal, 35*(1), 9-37.

Miles, R. E., and Snow, C. C. (1978). *Organizational strategy, structure, and process.* New York: McGraw-Hill.

Miles, R. E., Snow, C. C., Meyer, A. D., and Coleman, J., Henry J. (1978). Organizational strategy, structure, and process. *Academy of Management Review, 3*(3), 546-562.

Miller, C. C., Burke, L. M., and Glick, W. H. (1998). Cognitive diversity among upper-echelon executives: Implications for strategic decision processes. *Strategic Management Journal, 19*(1), 39.

Miller, C. C., and Cardinal, L. B. (1994). Strategic planning and firm performance: A synthesis of more than two decades of research. *Academy of Management Journal, 37*(6), 1649-1665.

Miller, D. (1983). The correlates of entrepreneurship in three types of firms. *Management Science, 29*(7), 770-791.

Miller, D. (1991). Stale in the saddle: CEO tenure and the match between organization and environment. *Management Science, 37*(1), 34-52.

Miller, D. (1993). The architecture of simplicity. *Academy of Management Review, 18*(1), 116-138.

Miller, D., and Friesen, P. H. (1978). Archetypes of strategy formulation. *Management Science, 24*(9), 921-933.

Miller, D., and Friesen, P. H. (1982). Innovation in conservative and entrepreneurial firms: Two models of strategic momentum. *Strategic Management Journal, 3*(1), 1-25.

Miller, D., and Friesen, P. H. (1983). Strategy-making and environment: The third link. *Strategic Management Journal, 4*(3), 221-235.

Miller, D., and Friesen, P. H. (1984). A longitudinal study of the corporate life cycle. *Management Science, 30*(10), 1161-1183.

Miller, D., and Toulouse, J. (1986). Strategy, structure, CEO personality and performance in small firms. *American Journal of Small Business, 10*(3), 47-62.

Milliken, F. J. (1987). Three types of perceived uncertainty about the environment: State, effect, and response uncertainty. *Academy of Management Review, 12*(1), 133-143.

Mintzberg, H. (1973). Strategy-making in three modes. *California Management Review, 16*(2), 44-53.

Mintzberg, H. (1978). Patterns in strategy formation. *Management Science, 24*(9), 934-948.

Mintzberg, H. (1991). Learning 1, planning 0 reply to Igor Ansoff. *Strategic Management Journal, 12*(6), 463-466.

Mintzberg, H. (1994a). Rethinking strategic planning part II: New roles for planners. *Long Range Planning, 27*(3), 22-30.

Mintzberg, H. (1994b). *The rise and fall of strategic planning: Reconceiving roles for planning, plans, planners.* New York: Free Press.

Mintzberg, H., and Lampel, J. (1999). Reflecting on the strategy process. *Sloan Management Review, 40*(3), 21-30.

Mintzberg, H., Raisinghani, D., and Théorêt, A. (1976). The structure of 'unstructured' decision processes. *Administrative Science Quarterly, 21*(2), 246-275.

Mintzberg, H., and Waters, J. A. (1985). Of strategies, deliberate and emergent. *Strategic Management Journal, 6*(3), 257-272.

Mitchell, R. K., Busenitz, L. W., Bird, B., Marie Gaglio, C., McMullen, J. S., Morse, E. A., et al. (2007). The central question in entrepreneurial cognition research 2007. *Entrepreneurship: Theory & Practice, 31*(1), 1-27.

Molloy, S., and Schwenk, C. R. (1995). The effects of information technology on strategic decision making. *Journal of Management Studies, 32*(3), 283-311.

Montgomery, M., Johnson, T., and Faisal, S. (2005). What kind of capital do you need to start a business: Financial or human? *The Quarterly Review of Economics and Finance, 45*(1), 103-122.

Moray, N., and Clarysse, B. (2005). Institutional change and resource endowments to science-based entrepreneurial firms. *Research Policy, 34*(7), 1010-1027.

Morgan, R. E., and Strong, C. A. (1998). Market orientation and dimensions of strategic orientation. *European Journal of Marketing, 32*(11/12), 1051-1973.

Morgan, R. E., and Strong, C. A. (2003). Business performance and dimensions of strategic orientation. *Journal of Business Research, 56*(3), 163-176.

Mosey, S., and Wright, M. (2007). From human capital to social capital: A longitudinal study of technology-based academic entrepreneurs. *Entrepreneurship: Theory & Practice, 31*(6), 909-935.

Mowery, D. C., Nelson, R. R., Sampat, B. N., and Ziedonis, A. A. (2001). The growth of patenting and licensing by U.S. universities: An assessment of the effects of the Bayh–Dole act of 1980. *Research Policy, 30*(1), 99-119.

Murray, A. I. (1989). Top management group heterogeneity and firm performance. *Strategic Management Journal, 10*, 125-141.

Mustar, P. (1997). How French academics create hi-tech companies: The conditions for success or failure. *Science and Public Policy, 24*(1), 37-43.

Mustar, P., Renault, M., Colombo, M. G., Piva, E., Fontes, M., Lockett, A., et al. (2006). Conceptualising the heterogeneity of research-based spin-offs: A multi-dimensional taxonomy. *Research Policy, 35*(2), 289-308.

Naldi, L., Nordqvist, M., Sjöberg, K., and Wiklund, J. (2007). Entrepreneurial orientation, risk taking, and performance in family firms. *Family Business Review, 20*(1), 33-47.

Nancy S., D. (1983). Route 128: The development of a regional high technology economy. *Research Policy, 12*(6), 299-316.

Narayanan, V. K., and Fahey, L. (1982). The micro-politics of strategy formulation. *Academy of Management Review, 7*(1), 25-34.

Ndonzuau, F. N., Pirnay, F., and Surlemont, B. (2002). A stage model of academic spin-off creation. *Technovation, 22*(5), 281-289.

Nerkar, A., and Shane, S. (2003). When do start-ups that exploit patented academic knowledge survive? *International Journal of Industrial Organization, 21*(9), 1391-1410.

Nicolaou, N., and Birley, S. (2003). Academic networks in a trichotomous categorisation of university spinouts. *Journal of Business Venturing, 18*(3), 333-359.

Nielsen, S. (2010). Top management team diversity: A review of theories and methodologies. *International Journal of Management Reviews, 12*(3), 301-316.

Noda, T., and Bower, J. L. (1996). Strategy making as iterated processes of resource allocation. *Strategic Management Journal, 17*, 159-192.

Nunnally, J. C. (1978). *Psychometric theory* (2nd Edition ed.). New York.

Nutt, P. C. (1998). How decision makers evaluate alternatives and the influence of complexity. *Management Science, 44*(8), 1148-1166.

O'Reilly, C. A., Snyder, R. C., and Boothe, J. N. (1993). Executive team demography and organizational change. In G. P. Huber, and W. H. Glick (Eds.), *Organizational change and redesign: Ideas and insights for improving performance* (pp. 147-175). New York: Oxford University Press.

Olofsson, C., and Wahlbin, C. (1984). Technology-based new ventures from technical universities: A Swedish case. *Proceedings of the 1984 Frontiers of Entrepreneurship Research Conference, Babson College and Georgia Institute of Technology*, 175.

O'Shea, R., Allen, T. J., O'Gorman, C., and Roche, F. (2004). Universities and technology transfer: A review of academic entrepreneurship literature. *Irish Journal of Management, 25*(2), 11-29.

Patzelt, H., zu Knyphausen-Aufseß, D., and Fischer, H. T. (2009). Upper echelons and portfolio strategies of venture capital firms. *Journal of Business Venturing, 24*(6), 558-572.

Pearce II, J. A., Freeman, E. B., and Robinson Jr., R. B. (1987). The tenuous link between formal strategic planning and financial performance. *Academy of Management Review, 12*(4), 658-675.

Pearce II, J. A., and Robbins, D. K. (1987). The impact of grand strategy and planning formality on financial performance. *Strategic Management Journal, 8*(2), 125-134.

Pearce II, J. A., Kramer, T. R., and Robbins, D. K. (1997). Effects of managers' entrepreneurial behavior on subordinates. *Journal of Business Venturing, 12*(2), 147-160.

Penrose, E. G. (1959). *The theory of the growth of the firm*. New York: Wiley.

Perry, J. T., Chandler, G. N., and Markova, G. (2011). Entrepreneurial effectuation: A review and suggestions for future research. *Entrepreneurship Theory and Practice, Special Issue*, 1-25.

Peter, S. I. (1997). *Kundenbindung als Marketingziel: Identifikation und Analyse zentraler Determinanten*. Wiesbaden.

Peteraf, M. A. (1993). The cornerstones of competitive advantage: A resource-based view. *Strategic Management Journal, 14*(3), 179-191.

Pettigrew, A. M. (1973). *Politics of organizational decision-making*. London: Tavistock.

Pettigrew, A. M. (1985). *The awakening giant; continuity and change in ICI*. Oxford: Basil Blackwell.

Pettigrew, A. M. (1992). On studying managerial elites. *Strategic Management Journal, 13*, 163-182.

Pfeffer, J. (1983). Organizational demography. In L. L. Cummings, and B. Staw (Eds.), *Research in organizational behaviour, Vol. 5* (pp. 299-357). Greenwich, CT: JAI Press.

Pirnay, F., Surlemont, B., and Nlemvo, F. (2003). Toward a typology of university spin-offs. *Small Business Economics, 21*(4), 355-369.

Podsakoff, N. P., Shen, W., and Podsakoff, P. M. (2006). In Ketchen D.J., and Bergh D.D. (Eds.), *The role of formative measurement models in strategic management research: Review, critique, and implications for future research*

Porac, J. F., and Thomas, H. (1990). Taxonomic mental models in competitor definition. *Academy of Management Review, 15*(2), 224-240.

Porter, M. E. (1980). *Competitive strategy: Techniques for analyzing industries and competitors*. New York: FreePress.

Porter, M. E. (1987). From competitive advantage to corporate strategy. *Harvard Business Review, 65*(3), 43-59.

Powell, T. C. (2001). Complete advantage: Logical and philosophical considerations. *Strategic Management Journal, 22*(9), 875.

Preisendorfer, P., and Voss, T. (1990). Organizational mortality of small firms: The effects of entrepreneurial age and human capital. *Organization Studies, 11*(1), 107-129.

Priem, R. L. (1990). Top management team group factors, consensus, and firm performance. *Strategic Management Journal, 11*(6), 469-478.

Priem, R. L., and Butler, J. E. (2001a). Is the resource-based 'view" a useful perspective for strategic management research? *Academy of Management Review, 26*(1), 22.

Priem, R. L., and Butler, J. E. (2001b). Tautology in the resource-based view and the implications of externally determined resource value: Further comments. *Academy of Management Review, 26*(1), 57.

Priem, R. L., Rasheed, A. M. A., and Kotulic, A. G. (1995). Rationality in strategic decision processes, environmental dynamism and firm performance. *Journal of Management, 21*(5), 913-929.

Prüfer, P., and Rexroth, M. (1996). Verfahren zur Evaluation von Survey-Fragen: Ein Überblick. *ZUMA-Nachrichten, 39*, 95-116.

Pugh, D. S., Hickson, D. J., Hinings, C. R., and Turner, C. (1968). Dimensions of organization structure. *Administrative Science Quarterly, 13*(1), 65-105.

Quinn, J. B. (1980). *Strategies for change: Logical incrementalism.* Homewood, IL: Dow-Jones-Irwin.

Radosevich, R. (1995). A model for entrepreneurial spin-offs from public technology sources. *International Journal of Technology Management, 10*(7/8), 879-893.

Rajagopalan, N., Rasheed, A. M. A., and Datta, D. K. (1993). Strategic decision processes: Critical review and future directions. *Journal of Management, 19*(2), 349-384.

Rappert, B., Webster, A., and Charles, D. (1999). Making sense of diversity and reluctance: Academic-industrial relations and intellectual property. *Research Policy, 28*(8), 873-890.

Rauch, A., Wiklund, J., Lumpkin, G. T., and Frese, M. (2009). Entrepreneurial orientation and business performance: An assessment of past research and suggestions for the future. *Entrepreneurship: Theory and Practice, 33*(3), 761-787.

Read, S., and Sarasvathy, S. D. (2005). Knowing what to do and doing what you know: Effectuation as a form of entrepreneurial expertise. *Journal of Private Equity, 9*(1), 45-62.

Read, S., Song, M., and Smit, W. (2009). A meta-analytic review of effectuation and venture performance. *Journal of Business Venturing, 24*(6), 573-587.

Rhenman, E. (1973). *Organization theory for long-range planning.* London: Wiley.

Richard, W. (1996). Strategy as practice. *Long Range Planning, 29*(5), 731-735.

Riesenhuber, F. (2008). *Technologiebasierte Chancen und Wachstum akademischer Spin-offs.* Wiesbaden: Gabler Edition Wissenschaft.

Rindova, V. P. (1999). What corporate boards have to do with strategy: A cognitive perspective. *Journal of Management Studies, 36*(7), 953-975.

Rindova, V. P., and Fombrun, C. J. (1999). Constructing competitive advantage: The role of firm-constituent interactions. *Strategic Management Journal, 20*(8), 691-710.

Roberts, E. B. (1991a). *Entrepreneurs in high technology - lessons from MIT and beyond.* New York: Oxford University Press.

Roberts, E. B. (1991b). The technological base of the new enterprise. *Research Policy, 20*(4), 283-298.

Roberts, E. B., and Malone, D. E. (1996). Policies and structures for spinning off new companies from research and development organizations. *R and D Management, 26*(1), 17-48.

Roberts, E. B., and Hauptman, O. (1986). The process of technology transfer to the new biomedical and pharmaceutical firm. *Research Policy, 15*(3), 107-119.

Robinson, J., Richard B., Salem, M. Y., Logan, J. E., and Pearce, I., J. (1986). Planning activities related to independent retail firm performance. *American Journal of Small Business, 11*(1), 19-26.

Robinson, P. B., and Sexton, E. A. (1994). The effect of education and experience on self-employment success. *Journal of Business Venturing, 9*(2), 141-156.

Rogers, E. M., and Hall, B. (1999). Technology transfer from university-based research centers. *Journal of Higher Education, 70*(6), 687-705.

Rothaermel, F. T., Agung, S. D., and Jiang, L. (2007). University entrepreneurship: A taxonomy of the literature. *Industrial and Corporate Change, 16*(4), 691-791.

Roure, J. B., and Keeley, R. H. (1990). Predictors of success in new technology based ventures. *Journal of Business Venturing, 5*(4), 201-220.

Rumelt, R. P. (1974). *Strategy, structure and economic performance.* Cambridge: Harvard University Press.

Sankaran, V. (2004). Regional transformation through technological entrepreneurship. *Journal of Business Venturing, 19*(1), 153-167.

Sapienza, H. J., Smith, K. G., and Gannon, M. J. (1988). Using subjective evaluations of organizational performance in small business research. *American Journal of Small Business, 12*(3), 45-53.

Sarasvathy, S. D. (2001). Causation and effectuation: Toward a theoretical shift from economic inevitability to entrepreneurial contingency. *Academy of Management Review, 26*(2), 243-263.

Schnell, R., Hill, P. B., and Esser, E. (2008). *Methoden der empirischen Sozialforschung.* München: Oldenbourg Wissenschaftsverlag.

Schoemaker, P. J. H. (2002). *Profiting from uncertainty: Stratgies for succeeding no matter what the future brings.* New York: Free Press.

Scholten, V. E. (2006). *The early growth of academic spin-offs: Factors influencing the early growth of Dutch spin-offs in the life sciences, ICT and consulting.* Wageningen University and Researchcentrum, the Netherlands.

Schröder, R. (2008). *Strategische Orientierungen für junge Technologieunternehmen.* Wiesbaden: Gabler Edition Wissenschaft.

Schumann, S. (2006). *Repräsentative Umfrage: Praxisorientierte Einführung in empirische Methoden und statistische Analyseverfahren.* München: Oldenbourg Wissenschaftsverlag.

Schumpeter, J. A. (1934). *The theory of economic development.* Cambridge, Massachusetts: Harvard University Press.

Schwenk, C. B., and Shrader, C. B. (1993). Effects of formal strategic planning on financial performance in small firms: A meta-analysis. *Entrepreneurship: Theory & Practice, 17*(3), 53-64.

Schwenk, C. H. (1986). Information, cognitive biases, and commitment to a course of action. *Academy of Management Review, 11*(2), 298-310.

Schwenk, C. R. (1984). Cognitive simplification processes in strategic decision-making. *Strategic Management Journal, 5*(2), 111-128.

Schwenk, C. R. (1995). Strategic decision making. *Journal of Management, 21*(3), 471.

Selznick, P. (1957). *Leadership in administration.* New York: Harper & Row.

Sexton, D. L., and Bowman, N. (1985). The entrepreneur: A capable executive and more. *Journal of Business Venturing, 1*(1), 129-140.

Sexton, D. L., and Van Auken, P. (1985). A longitudinal study of small business strategic planning. *Journal of Small Business Management, 23*(1), 7-15.

Shane, S. (2000). Prior knowledge and the discovery of entrepreneurial opportunities. *Organization Science, 11*(4), 448.

Shane, S. (2002). Selling university technology: Patterns from MIT. *Management Science, 48*(1), 122-137.

Shane, S. (2004). *Academic entrepreneurship: University spinoffs and wealth creation.* Cheltenham, Northampton: Edward Elgar.

Shane, S., and Stuart, T. (2002). Organizational endowments and the performance of university start-ups. *Management Science, 48*(1), 154-170.

Shane, S., and Venkataraman, S. (2000). The promise of entrepreneurship as a field of research. *Academy of Management Review, 25*(1), 217-226.

Sharfman, M. P., and Dean, J. W. (1991). Conceptualizing and measuring the organizational environment: A multidimensional approach. *Journal of Management, 17*(4), 681-700.

Shepherd, D. A., Zacharakis, A., and Baron, R. A. (2003). VCs' decision processes: Evidence suggesting more experience may not always be better. *Journal of Business Venturing, 18*(3), 381-401.

Shrader, C. B., Taylor, L., and Dalton, D. R. (1984). Strategic planning and organizational performance: A critical appraisal. *Journal of Management, 10*(2), 149-171.

Shrader, R., and Siegel, D. S. (2007). Assessing the relationship between human capital and firm performance: Evidence from technology-based new ventures. *Entrepreneurship: Theory & Practice, 31*(6), 893-908.

Shrivastava, P., and Grant, J. H. (1985). Empirically derived models of strategic decision-making processes. *Strategic Management Journal, 6*(2), 97-113.

Slevin, D. P., and Covin, J. G. (1997). Strategy formation patterns, performance, and the significance of context. *Journal of Management, 23*(2), 189-209.

Smilor, R. W., Gibson, D. V., and Dietrich, G. B. (1990). University spin-out companies: Technology start-ups from UT-Austin. *Journal of Business Venturing, 5*(1), 63-76.

Sminia, H. (1994). *Turning the wheels of change.* Groningen: Wolters-Noordhoff.

Sminia, H. (2009). Process research in strategy formation: Theory, methodology and relevance. *International Journal of Management Reviews, 11*(1), 97-125.

Smith, J. A. (1998). Strategies for start-ups. *Long Range Planning, 31*(6), 857-872.

Smith, K. G., Smith, K. A., Sims Jr., H. P., O'Bannon, D. P., Scully, J. A., and Olian, J. D. (1994). Top management team demography and process: The role of social integration and communication. *Administrative Science Quarterly, 39*(3), 412-438.

Song, M., Podoynitsyna, K., van der Bij, H., and Halman, J. I. M. (2008). Success factors in new ventures: A meta-analysis. *Journal of Product Innovation Management, 25*(1), 7-27.

Spence, M. T., and Brucks, M. (1997). The moderating effects of problem characteristics on experts' and novices' judgments. *Journal of Marketing Research (JMR), 34*(2), 233-247.

Stankiewicz, R. (1994). University firms: Spin-off companies from universities. *Science and Public Policy, 21*, 99-107.

Starr, J., and Bygrave, W. D. (1991). The assets and liabilities of prior start-up experience: An exploratory study of multiple venture entrepreneurs. In N. C. Churchill, et al. (Eds.), *Frontiers of entrepreneurship research 1991* (pp. 213-227). Wellesley, MA: Babson College.

Steffensen, M., Rogers, E. M., and Speakman, K. (2000). Spin-offs from research centers at a research university. *Journal of Business Venturing, 15*(1), 93-111.

Steinkühler, R. H. (1994). *Technologiezentren und Erfolg von Unternehmensgründungen.* Wiesbaden: Deutscher Universitätsverlag.

Stevenson, H. H., and Jarillo, J. C. (1990). A paradigm of entrepreneurship: Entrepreneurial management. *Strategic Management Journal, 11*(4), 17-27.

Stinchcombe, A. L. (1965). Social structures and organizations. In James G. March (Ed.), *Handbook of organizations* (pp. 142-193). Chicago: Rand-McNally.

Stuart, R. W., and Abetti, P. A. (1990). Impact of entrepreneurial and management experience on early performance. *Journal of Business Venturing, 5*(3), 151-162.

Szyperski, N., and Klandt, H. (1981). *Wissenschaftlich-technische Mitarbeiter von Forschungs- und Entwicklungseinrichtungen als potentielle Spin-off Gründer: Eine empirische Studie zu den Entstehungsfaktoren von innovativen Unternehmensgründungen im Lande Nordrhein-Westfalen.* Opladen: Westdeutscher Verlag.

Teece, D. J., Pisano, G., and Shuen, A. (1997). Dynamic capabilities and strategic management. *Strategic Management Journal, 18*(7), 509-533.

Tellis, G. J., and Golder, P. N. (2002). *Will and vision: How latecomers grow to dominate markets.* New York: McGraw-Hill.

Thomas, A. S., Litschert, R. J., and Ramaswamy, K. (1991). The performance impact of strategy - manager coalignment: An empirical examination. *Strategic Management Journal, 12*(7), 509-522.

Torben Juul, A. (2000). Strategic planning, autonomous actions and corporate performance. *Long Range Planning, 33*(2), 184-200.

Tushman, M. L., and Romanelli, E. (1985). Organizational evolution: A metamorphosis model of convergence and reorientation. *Research in Organizational Behavior, 7*, 171-222.

Tushman, M. L. (1977). A political approach to organizations: A review and rationale. *Academy of Management Review, 2*(2), 206-216.

Tyler, B. B., and Steensma, H. K. (1998). The effects of executives' experiences and perceptions on their assessment of potential. *Strategic Management Journal, 19*(10), 939.

Ucbasaran, D., Westhead, P., and Wright, M. (2006). *Habitual entrepreneurs.* Aldershot, UK: Edward Elgar.

Ucbasaran, D., Alsos, G. A., Westhead, P., and Wright, M. (2008). *Habitual entrepreneurs.* Boston - Delft.

Ucbasaran, D., Lockett, A., Wright, M., and Westhead, P. (2003). Entrepreneurial founder teams: Factors associated with member entry and exit. *Entrepreneurship: Theory & Practice, 28*(2), 107-127.

Ucbasaran, D., Westhead, P., and Wright, M. (2009). The extent and nature of opportunity identification by experienced entrepreneurs. *Journal of Business Venturing, 24*(2), 99-115.

Ucbasaran, D., Westhead, P., Wright, M., and Flores, M. (2010). The nature of entrepreneurial experience, business failure and comparative optimism. *Journal of Business Venturing, 25*(6), 541-555.

Unger, J. M., Rauch, A., Frese, M., and Rosenbusch, N. (2011). Human capital and entrepreneurial success: A meta-analytical review. *Journal of Business Venturing, 26*(3), 341-358.

Unterkofler, G. (1989). *Erfolgsfaktoren innovativer Unternehmensgründungen - ein gestaltungsorientierter Lösungsansatz betriebswirtschaftlicher Gründungsprobleme.* Frankfurt am Main: Lang.

Van de Ven, A. H., Hudson, R., and Schroder, D. M. (1984). Designing new business start-Up's entrepreneurial, organizational, and ecological considerations. *Journal of Management, 10*, 87-107.

Van Dierdonck, R., and Debackere, K. (1988). Academic entrepreneurship at Belgian universities. *R&D Management, 18*(4), 341-353.

Van Looy, B., Callaert, J., and Debackere, K. (2006). Publication and patent behavior of academic researchers: Conflicting, reinforcing or merely co-existing? *Research Policy, 35*(4), 596-608.

Vanaelst, I., Clarysse, B., Wright, M., Lockett, A., Moray, N., and S'Jegers, R. (2006). Entrepreneurial team development in academic spinouts: An examination of team heterogeneity. *Entrepreneurship: Theory & Practice, 30*(2), 249-271.

Venkataraman, S., and Sarasvathy, S. D. (2001). Strategy and entrepreneurship: Outlines of an untold story. In M. A. Hitt, E. Freeman and J. S. Harrison (Eds.), *Handbook of strategic management* (pp. 650-668). Oxford: Blackwell.

Venkatraman, N. (1989). Strategic orientation of business enterprises: The construct, dimensionality, and measurement. *Management Science, 35*(8), 942-962.

Venkatraman, N., and Grant, J. H. (1986). Construct measurement in organizational strategy research: A critique and proposal. *The Academy of Management Review, 11*(1), 71-87.

Venkatraman, N., and Ramanujam, V. (1986). Measurement of business performance in strategy research: A comparison of approaches. *Academy of Management Review, 11*(4), 801-814.

Virany, B., and Tushman, M. L. (1986). Top management teams and corporate success in an emerging industry. *Journal of Business Venturing, 1*(3), 261-274.

Vohora, A., Wright, M., and Lockett, A. (2004). Critical junctures in the development of university high-tech spinout companies. *Research Policy, 33*(1), 147-175.

Wally, S., and Baum, J. R. (1994). Personal and structural determinants of the pace of strategic decision making. *Academy of Management Journal, 37*(4), 932-956.

Walter, A., Auer, M., and Ritter, T. (2006). The impact of network capabilities and entrepreneurial orientation on university spin-off performance. *Journal of Business Venturing, 21*(4), 541-567.

Watson, T. J. (1995). Entrepreneurship and professional management: A fatal distinction. *International Small Business Journal, 13*(2), 34-46.

Watson, W. E., Ponthieu, L. D., and Critelli, J. W. (1995). Team interpersonal process effectiveness in venture partnerships and its connection to perceived success. *Journal of Business Venturing, 10*(5), 393-411.

Weatherston, J. (1995). Academic entrepreneurs: Is a spin-off company too risky? *Proceedings of the 40th International Council on Small Business, 18-21 June,* Sydney

Weinzimmer, L. G. (1997). Top management team correlates of organizational growth in a small business context: A comparative study. *Journal of Small Business Management, 35*(3), 1-10.

Wernerfelt, B. (1984). A resource-based view of the firm. *Strategic Management Journal, 5*(2), 171-180.

Westhead, P., and Storey, D. J. (1995). Links between higher education institutions and high technology firms. *Omega, 23*(4), 345-360.

Westhead, P., Ucbasaran, D., and Wright, M. (2005). Decisions, actions, and performance: Do novice, serial, and portfolio entrepreneurs differ? *Journal of Small Business Management, 43*(4), 393-417.

Westhead, P., and Wright, M. (1998). Novice, portfolio, and serial founders: Are they different? *Journal of Business Venturing, 13*(3), 173-204.

Wiersema, M. F., and Bantel, K. A. (1992). Top management team demography and corporate strategic change. *Academy of Management Journal, 35*(1), 91-121.

Wiklund, J., and Shepherd, D. (2003). Knowledge-based resources, entrepreneurial orientation, and the performance of small and medium-sized businesses. *Strategic Management Journal, 24*(13), 1307-1314.

Wiklund, J., and Shepherd, D. (2005). Entrepreneurial orientation and small business performance: A configurational approach. *Journal of Business Venturing, 20*(1), 71-91.

Willard, G. E., Krueger, D. A., and Feeser, H. R. (1992). In order to grow, must the founder go: A comparison of performance between founder and non-founder managed high-growth manufacturing firms. *Journal of Business Venturing, 7*(3), 181-194.

Williams, K. Y., and O'Reilly, C. A. (1998). Demography and diversity in organizations: A review of 40 years of research. In B. M. Staw, and L. Cummings (Eds.), *Research in organizational behaviour: An annual series of analytical essays and critical reviews* (pp. 77-140). Greenwich, CT: JAI Press.

Wiltbank, R., Dew, N., Read, S., and Sarasvathy, S. D. (2006). What to do next? The case for non-predictive strategy. *Strategic Management Journal, 27*(10), 981-998.

Wright, M., Clarysse, B., Mustar, P., and Lockett, A. (2007). *Academic entrepreneurship in Europe*. Cheltenham: Edward Elgar Publishing.

Wright, M., Birley, S., and Mosey, S. (2004). Entrepreneurship and university technology transfer. *Journal of Technology Transfer, 29*(3-4), 235-246.

Wright, M., Hmieleski, K. M., Siegel, D. S., and Ensley, M. D. (2007). The role of human capital in technological entrepreneurship. *Entrepreneurship: Theory & Practice, 31*(6), 791-806.

Wright, M., Lockett, A., Clarysse, B., and Binks, M. (2006). University spin-out companies and venture capital. *Research Policy, 35*(4), 481-501.

Wright, M., Vohora, A., and Lockett, A. (2004). The formation of high-tech university spinouts: The role of joint ventures and venture capital investors. *Journal of Technology Transfer, 29*(3-4), 287-310.

Yasai-Ardekani, M., and Nystrom, P. C. (1996). Designs for environmental scanning systems: Tests of a contingency theory. *Management Science, 42*(2), 187-204.

Yusof, M., and Jain, K. K. (2010). Categories of university-level entrepreneurship: A literature survey. *International Entrepreneurship Management Journal, 6*(1), 81-96.

Zahra, S. A. (1993). Environment, corporate entrepreneurship, and financial performance: A taxonomic approach. *Journal of Business Venturing, 8*(4), 319-340.

Zahra, S. A. (1996). Technology strategy and financial performance: Examining the moderating role of the firm's competitive environment. *Journal of Business Venturing, 11*(3), 189-219.

Zahra, S., and Dess, G. G. (2001). Entrepreneurship as a field of research: Encouraging dialogue and debate. *Academy of Management Review, 26*(1), 8-10.

Zahra, S. A., and Covin, J. G. (1993). Business strategy, technology policy and firm performance. *Strategic Management Journal, 14*(6), 451-478.

Zaltman, G., Duncan, R., and Holbek, J. (1973). *Innovations and organizations.* New York: Wiley.

Zinnbauer, M., and Eberl, M. (2005). Überprüfung der Spezifikation und Güte von Strukturgleichungsmodellen. *Wirtschaftswissenschaftliches Studium - Zeitschrift für Ausbildung und Hochschulkontakt, 34*(10), 566-572.

Zucker, L. G., Darby, M. R., and Armstrong, J. S. (2002). Commercializing knowledge: University science, knowledge capture, and firm performance in biotechnology. *Management Science, 48*(1), 138-153.